South-East Asian Social Science Monographs

Muslim Separatism:
The Moros of Southern Philippines
and the Malays of Southern Thailand

Muslim Separatism:
The Moros of Southern Philippines and the Malays of Southern Thailand

W. K. Che Man

SINGAPORE
OXFORD UNIVERSITY PRESS
OXFORD NEW YORK
1990

Oxford University Press
Oxford New York Toronto
Delhi Bombay Calcutta Madras Karachi
Petaling Jaya Singapore Hong Kong Tokyo
Nairobi Dar es Salaam Cape Town
Melbourne Auckland
and associated companies in
Berlin Ibadan

Oxford is a trade mark of Oxford University Press

© Oxford University Press Pte. Ltd. 1990

Published in the United States by
Oxford University Press Inc., New York

All rights reserved. No part of this publication may be reproduced,
stored in a retrieval system, or transmitted, in any form or by any means,
electronic, mechanical, photocopying, recording or otherwise,
without the prior permission of Oxford University Press

ISBN 0 19 588924 X

British Library Cataloguing in Publication Data
Che Man, W.K. (Wan Kadir, 1946–)
Muslim separatism: the Moros of Southern Philippines
and the Malays of Southern Thailand.—(South-East Asian
Social Science monographs)
1. Asia—East Asia. Separatist movements, Muslim
I. Title II. Series
322.4'2'0951

ISBN 0-19-588924-X

Library of Congress Cataloging-in-Publication Data
K. Che Man (Kadir Che Man), W., 1946–
Muslim separatism: the Moros of southern Philippines and the
Malays of southern Thailand/W.K. Che Man.
p. cm.—(South-East Asian social science monographs)
ISBN 0-19-588924-X (U.S.)
1. Mindanao Island (Philippines)—History—Autonomy and
independence movements. 2. Muslims—Philippines—Mindanao Island—
Politics and government. 3. Pattani (Thailand: Province)—
History—Autonomy and independence movements. 4. Muslims—
Thailand—Pattani (Province)—Politics and government. 5. Malays
(Asian people)—Thailand—Pattani (Province)—Politics and
government. I. Title. II. Series.
DS688.M2K2 1990
959.9'7—dc20
89–36289
CIP

Printed in Malaysia by Peter Chong Printers Sdn. Bhd.
Published by Oxford University Press Pte. Ltd.,
Unit 221, Ubi Avenue 4, Singapore 1440

*To the memory of my father, Hajji Daud,
and to my mother, Hajjah Wan Zainab,
my dedicated teachers*

It should be plain that separatism, generating the collective will and organization to go it alone, is more than mere nostalgia or emphasis on ethnicity. Going back and gathering with one's own kind is only part of such closing: More important is a synergy from within, a surge of collective vitality that enables the group to create, develop, and carry on with a new sense of togetherness, which is a loss for the larger group but a gain for the smaller one—when successful.

Orrin E. Klapp
(Quoted in Williams, 1982)

Foreword

IN recent years there has been something of a resurgence of ethnicity and of ethnic nationalism or subnationalism, not only in the post-colonial states of Africa, Asia, and the Pacific, but also in the developed states of the West, and in Eastern Europe and Soviet Asia. Such developments have confounded the predictions of those commentators on 'political development' in the 1960s and 1970s—both those who wrote from a 'modernization' perspective and those of a Marxist orientation—who generally saw ethnicity as a transitory phenomenon which would ultimately yield to the dominant forces of national integration or of class conflict. The persistence of ethnicity, the increasing incidence of ethnic conflict, and the frequent internationalization of such conflict have prompted a reassessment of ethnicity, from its role in the formation of nation-states to its significance as a source of domestic and international conflict and instability. Yet despite the salience of ethnic factors in the politics of the post-colonial states of Africa and Asia, where ethnic conflict has often been a seemingly inevitable consequence of arbitrarily drawn colonial boundaries, much of the growing literature on ethnicity is heavily Eurocentric.

This study, by a South-East Asian scholar himself shaped by the ethnicity of his geographical circumstances, helps to redress the balance. In it Dr W. K. Che Man examines ethnic nationalist movements among the Muslims of southern Philippines and southern Thailand. He traces the historical and cultural roots of the movements and examines the way in which Islam served as a unifying force, defining Moro and Malay ethnic identity. He also describes the domestic and external factors which served to stimulate the emergence of the separatist movements at a point of time and to influence the course of the separatist struggles in the two countries. On the basis of extensive field-work in both Thailand and the Philippines, Che Man analyses the internal dynamics of the two movements and relates this to the issue of leadership in Moro and Malay society.

The comparative study of the Moro and Patani movements suggests a number of parallels and a number of divergences. But above all, it illustrates how ethnicity can persist in the face of governmental repression and of internal fragmentation. In the Philippines, the Aquino government has recently approved measures to grant 'Muslim autonomy' in the south, but the circumstances under which the autonomy provisions have been formulated suggest that the government has learnt little from the experience of the past two decades.

Although there have been other studies of Muslim separatism in Thailand and the Philippines, Che Man's study is likely to remain something of a benchmark, not only for what it tells us about Thailand and the Philippines, but for the insights it offers for the broader study of ethnic separatism.

Canberra R. J. MAY
September 1989

Preface

THIS book is a slightly revised version of my doctoral dissertation submitted to the Australian National University, Canberra. The study examines ethnic separatist politics among the Muslim minority populations in Mindanao in the Philippines, and in Patani in Thailand. It addresses several questions. Having regard to the alternative explanations offered for the persistence and resurgence of ethnic nationalism, how does one account for the survival of Moro and Malay identity over hundreds of years in the face of repeated attempts to integrate these ethnic groups into the larger political entities of the Philippines and Thailand respectively? Under what conditions will an ethnic group choose to take action aimed at separatism as opposed to, for example, working for change within the system? Are separatist struggles essentially inspired by cultural conflict or by economic deprivation?

Beyond these broad questions, the book scrutinizes the form which the separatist movements have taken over time and looks in detail at the structure, ideology, and leadership of the struggles. An account is also given of Manila's and Bangkok's responses to their respective separatist problems. It is argued that attempts to solve the problem of ethnic separatism through either socio-economic programmes or military action are doomed to failure. It is thus hoped that this modest exercise will throw some light on general phenomena of ethnic separatism and, in doing so, help to provide an understanding of the long-term prospects for separatist movements.

I wish to express my gratitude to the Australian National University for the award of a Ph.D. scholarship and research funds to conduct field-work in the Philippines and Thailand. I also thank the Muslim World League, Mecca, and the Royal Embassy of the Kingdom of Saudi Arabia in Canberra for making possible a trip to the Kingdom.

I am indebted to my supervisors, Dr R. J. May, Professor J. A. C. Mackie, Dr James P. Piscatori, and Dr John S. L. Girling, who played a significant role in shaping my dissertation. Special thanks must be conveyed to Dr Francisco Nemenzo for his valuable academic assistance and for providing me with access to certain contacts in the Philippines and to Dr Harold A. Crouch for his useful suggestions and comments. Gratitude is due, too, to Dr William J. O'Malley, Dr Kevin Hewison, Dr Mohamed Yusoff Ismail, Dr Radin Fernando, and Ms Linda Allen who, during my years at the ANU, gave me much assistance, both academically and otherwise.

Appreciation is acknowledged to Dr Wilfredo Arce (Director), Dr Jean Miralao, and members of the Institute of Philippine Culture, Ateneo de Manila University, for allowing me to be affiliated with the institution during the course of my field-work. Similar acknowledgement is expressed to Mr Moctar Matuan (Director) and members of the Peter Gowing Memorial Research Center and to Mr Aziz Tawagon (Director) and members of the Dansalan Extension Services at the Dansalan College Foundation, where my major library research was done.

My greatest debt, however, is to leaders and members of the different fronts in Mindanao and in Patani. They not only permitted me to live with them and observe their activities, but also extended their hospitality and helping hands, which made my difficult tasks a lot easier. Some of their names deserve to be recorded here: in Mindanao, Commander Iqra, Commander Solitario, Commander Narrah, Commander Gadil, Ustaz Alim Aziz, Ustaz Muammar Qutb, and Ustaz Mohammad Ameen; in Patani, Ustaz Badri Hamdan, Ustaz Ibn Al-Walid Al-Khalid, Ustaz Karim Hassan, and Commander Che Muda.

During my field studies, I received much help, hospitality, and guidance from various individuals. In their own capacities, they certainly contributed to this work. In the Philippines, these names must be recorded: Mr Mario Mapanao, Mr Abdul Malik Solaiman, Mr Mubashar Abbas, Mr Al-Razzy, Mr Adab Gumpong, Ms Fatmawati Salapuddin, Mrs Princes Nemenzo, Mr Rody Rodil, Mr Paladan Badron, Mr Salic Abdul, Datu Micheal O. Mastura, Dr Ahmad (Jun) Alonto, Dr Nagasura T. Madale, Mr Pangalian Balindong, Hajji Omar M. Pasigan, and Mr Abu Khalil.

In Patani, most of the help and hospitality I received during my fieldwork came from my relatives and friends in the Muslim provinces. Most important of all, however, is my father-in-law, Hajji A. Rashid A. Rahman, a former MP and Mayor of Narathiwat town, who provided me with access to better contacts with government officials and Muslim leaders. In Saudi Arabia, I am greatly indebted to my old friends who worked hard to enable me to carry out my field study successfully.

Finally, I am ever grateful to my wife, Nafisah, who has been so understanding and helpful, as were our children, Faris, Daud, and Kasim. They have been my constant source of inspiration and emotional vigour; and without their help and encouragement, this work could not have been done.

Patani W. K. CHE MAN
December 1988

Contents

Foreword vii
Preface ix
Appendices xiii
Tables xiv
Figures xvi
Maps xvi
Abbreviations xvii

Introduction: Theoretical Considerations 1
The Persistence of Ethnicity 2

1 Moro and Malay Societies: Historical, Economic, and Social Backgrounds 17
The Mindanao and Patani Scenes 17
The Mindanao and Patani Scenes Compared 43

2 History of Muslim Resistance Movements in the Twentieth Century 46
The Moro Resistance Movement 46
The Malay Resistance Movement 62
The Moro and the Malay Resistance Movements Compared 70

3 Muslim Separatism: The Moros and the Malays 74
The Moros 74
The Malays 97
The Moro and the Malay Separatisms Compared 113

4 Leadership in the Moro and the Malay Societies 116
Leadership in Moro Society 116
Leadership in the Malay Society in Patani 129
Leadership in the Moro and the Malay Societies Compared 136

5 External Influences on and Government Responses to the Muslim Separatist Conflicts 138
The Philippines 138

Thailand	158
External Influences on and Government Responses to the Moro and the Malay Separatist Conflicts Compared	169
6 Conclusion	172
Glossary	180
Appendices	183
Bibliography	218
Index	234

Appendices

 1 The Tripoli Agreement 183
 2 The Manifesto of the Muslim Independence Movement 187
 3 The Manifesto of the Moro National Liberation Front 189
 4 Structure of the Moro Organizations 191
 5 The Nine-point Proposal of the Reformist Group 197
 6 Resolutions of the Reformist Group: Rejection of Misuari's Leadership and Acceptance of Autonomy 198
 7 The MORO Manifesto 200
 8 Structure of the Malay Organizations 209
 9 Resolution No. 18 of the Political Committee at the Fifth Islamic Conference of Foreign Ministers Held at Kuala Lumpur on June 21–25, 1974 213
10 Working Paper for the Meeting of the Ministerial Four-member Committee 215
11 Resolution No. 10 of the Political Committee at the Sixth Islamic Conference of Foreign Ministers 217

Tables

1.1 Population of Bangsa Moro Cultural–Linguistic Groups in Mindanao, 1980 and 1982 — 19
1.2 Estimated Moro and Non-Moro Populations in Mindanao, 1903–1980 — 25
1.3 Utilization of Cropland in Mindanao, 1977 (Export Crops) — 26
1.4 Utilization of Cropland in Mindanao, 1977 (Local Consumption) — 26
1.5 Mindanao Exports, 1977 — 27
1.6 Average Gross Regional Domestic Product and Per Capita Income in the Muslim Area (Regions IX and XII), in Mindanao, and in the Philippines (peso) — 28
1.7 Gross Provincial Product and Per Capita Income in the Patani Region, in the Southern Region, and in Thailand, 1980 — 36
1.8 Area and Population of the Muslim Provinces of Southern Thailand — 36
1.9 Provincial Administrative Units of the Muslim Provinces of Southern Thailand — 41
2.1 Number of Moro Students Awarded Scholarships by Muslim Countries, 1977–1981 — 58
2.2 Salary of a Moro Preacher by Islamic Call Centre — 59
2.3 Number of Registered Mosques, Religious Schools, Religious Teachers, and Students in the Patani Region — 69
3.1 Potential Muslim Supporters Claimed by MNLF-Reformist Group, 1984 — 91
3.2 Members of the Central Committee of the BNPP, as of 3 August 1986 — 103
3.3 Members of the Working Committee of the BNPP in Mecca Who Were Holding Portfolios in 1986 — 111
4.1 The Commission on National Integration Programme: Number of Scholars and Graduates, 1958–1967 — 118
4.2 Moro Leadership Group Combinations — 119
4.3 Leadership Group Distribution among Muslim Officials, Autonomous Region XII (per cent) — 119

4.4 Leadership Group Distribution among the Elected Muslim Leaders at the Provincial and Municipal Levels (per cent)	120
4.5 Leadership Group Distribution among the Elected and Appointed Muslim Leaders at the Regional, Provincial, and Municipal Levels (per cent)	120
4.6 Some Active Moro Leaders in the Early Movement, 1970–1972	127
4.7 Members of the Central Committee of the MNLF, 1974	128
4.8 Members of Parliament in the Four Muslim Provinces of Southern Thailand, 1986	134
5.1 Members of the OIC, 1985	143

Figures

A4.1	Structure of the MNLF Organization (A)	191
A4.2	Structure of the MNLF Organization (B)	192
A4.3	Structure of the MNLF Organization (C)	192
A4.4	Structure of the MNLF Organization (D)	193
A4.5	Structure of the BMLO Organization	193
A4.6	Structure of the MILF Organization (A)	194
A4.7	Structure of the MILF Organization (B)	195
A4.8	Structure of the MORO Organization	196
A8.1	Structure of the BNPP Organization (A)	209
A8.2	Structure of the BNPP Organization (B)	210
A8.3	Structure of the BBMP Organization	211
A8.4	Structure of the PULO Organization	212

Maps

1	South-East Asia	18
2	The Philippines	20
3	Thailand	33

Abbreviations

AFP	Armed Forces of the Philippines
ASEAN	Association of Southeast Asian Nations
BBMP	Barisan Bersatu Mujahideen Patani
BIPP	Barisan Islam Pembebasan Patani
BMA	Bangsa Moro Army
BMILO	Bangsa Muslimin Islamic Liberation Organization
BMLO	Bangsa Moro Liberation Organization
BNPP	Barisan Nasional Pembebasan Patani
BRN	Barisan Revolusi Nasional
CAMP	Consultative Assembly of the Moro People
CEMCOM	Central Mindanao Command
CHDF	Civilian Home Defense Forces
CM	Council for Mosque
CNI	Commission on National Integration
CPM	Communist Party of Malaya
CPP	Communist Party of the Philippines
CWC	Central Working Committee
EDCOR	Economic Development Corps
EDIPTAF	Educational Development Implementing Projects Task Force
EEC	European Economic Community
FAPE	Fund for Assistance to Private Education
GAMPAR	Gabungan Melayu Patani Raya
GPP	Gross Provincial Product
GRDP	Gross Regional Domestic Product
HPC	High Political Council
ICFM	Islamic Conference of Foreign Ministers
IDB	Islamic Development Bank
IDP	Islamic Directorate of the Philippines
IMF	International Monetary Fund
JOTAF	Jolo Task Force
KBL	Kilusang Bagong Lipunan (New Society Movement)
LOI	Letter of Instruction
MAP	Muslim Association of the Philippines
MDA	Mindanao Development Authority
MDF	Muslim Development Fund

MILF	Moro Islamic Liberation Front
MIM	Muslim Independence Movement
MINSUPALA	Mindanao–Sulu–Palawan
MMA	Ministry of Muslim Affairs
MNLF	Moro National Liberation Front
MORO	Moro Revolutionary Organization
MSU	Mindanao State University
MWC	Muslim World Congress
MWL	Muslim World League
MYNA	Moro Youth National Assembly
NAFP	New Armed Forces of the Philippines
NATO	North Atlantic Treaty Organization
NDF	National Democratic Front
NLAPP	National Liberation Army of the Patani People
NLF	National Liberation Front of South Vietnam
NMRC	Northern Mindanao Revolutionary Committee
NPA	New People's Army
NUSP	National Union of Students of the Philippines
OIC	Organization of the Islamic Conference
PAB	Philippine Amanah Bank
PAS	Partai Islam se-Malaysia
PC	Philippine Constabulary
PCIA	Provincial Council for Islamic Affairs
PD	Presidential Decree
PLO	Palestine Liberation Organization
PPM	Patani People's Movement
PSRM	Partai Sosialis Rakyat Malaya (Malaysia)
PSSD	Pilgrim's Special Saving Deposit
PTF-RDM	Presidential Task Force for Reconstruction and Development of Mindanao
PULO	Patani United Liberation Organization
QMC	Quadripartite Ministerial Commission
RAD	Rehabilitation and Development
RNRC	Ranao Norte Revolutionary Committee
RSRC	Ranao Sur Revolutionary Committee
SATAG	Sulu Air Task Group
SEC	Supreme Executive Council
SECCOM	Section Committee
SPARE	Special Programme of Assistance for Rehabilitation of Evacuees
SPDA	Southern Philippines Development Administration
SRC	Supreme Revolutionary Council
UIFO	Union of Islamic Forces and Organizations
UN	United Nations
UNCTAD	United Nations Commission on Trade and Development
USNO	United Sabah National Organization
WC	Working Committee

Introduction:
Theoretical Considerations

ETHNIC pluralism has occurred throughout mankind's history. Frequently, it has been a consequence of conquest. In an analysis of 132 contemporary sovereign states, Said and Simmons (1976: 10) suggest that only 12 can be considered ethnically homogeneous. One may, however, distinguish different types of ethnically heterogeneous society, depending on the socio-historical basis on which minority communities are differentiated.

Minority communities may be divided into three major categories. The first consists of minority immigrant populations. These are often the result of labour-recruiting policies during a colonial period; examples include the Blacks in the United States, the Chinese in Malaysia, and the Africans in Guyana. Such populations usually have no attachment to specific geographical areas being scattered throughout the territory of their new countries, and their sense of common identity is often weak. More significantly, such minorities have no real separatist alternative (for a discussion of the separatist option, see Anderson *et al.*, 1967: 67–74; Smith, 1982: 32–6); their predominant concern, indeed, is with acceptance within the larger society.

The second category, represented by groups such as the Aborigines in Australia, the American Indians in the United States, and the Innuits in Canada, constitutes indigenous peoples who, as a result of colonial settlement, have become a minority in their own land. In many cases, the impact of colonial settlement has not only reduced their populations but has driven them into isolation. Not surprisingly, some of the indigenous communities are fighting for their survival—a struggle which Darwin (1979: 75–94) has described as the course of natural selection, the survival of the fittest. But while demands for land rights and homelands may be regarded as a form of separatism, their real purpose is to limit outside encroachment on the traditional society.

The third category of ethnic minority results from incorporating hitherto autonomous peoples under an alien authority, or from diminishing the sphere of authority of local and regional groups under a policy of unification and centralization under a national government (Oberschall, 1973: 44). Examples include: the Basques in Spain; the French-speaking

Bernese in Switzerland; the South Tyroleans in Italy; the Bretons in France; the French in Quebec, Canada; the Walloon and Flemish peoples in Belgium; the Outer Islanders, including the East Timorese, in Indonesia; the Karens and the Shans in Burma; the Kashmir and the Naga peoples in India; the Tamils in Sri Lanka; the Tibetans and the Kazakhs in China, the Kurds in Iraq and Iran; the Armenians in Turkey and the USSR; the Eritreans in Ethiopia; the Saharawis in Morocco; the Ibos in Nigeria; the Moros in the Philippines; and the Malays in Thailand. Such groups normally identify strongly with a specific region and regard separation as a political option.

It is with this third category of minority that the book is concerned. The study focuses on the Moro and the Malay minority groups in the Philippines and Thailand respectively. Among these groups, separatism is perceived not only as having a socio-historical logic but also as being a practical possibility because of the coincidence of geography and cultural plurality, distance from the centre of authority, and the support of sympathetic groups outside the state.

The Persistence of Ethnicity

The ethnic community or *ethnie* has emerged and re-emerged at different periods since the early third millennium BC, and has remained as a socio-cultural 'model' for human organization and communication to the present day (Smith, 1987: 32). An ethnic community may be defined as 'a self-perceived group of people who hold in common a set of traditions not shared by the others with whom they are in contact' (De Vos and Romanucci-Ross, 1975: 9; Raymond L. Hall, 1979: xx). This definition assumes that for an ethnic group to emerge, there must be some 'primordial' ties around which to build a sense of community. These primordial ties include such traditions as common myths of descent or place of origin, sense of historical continuity, and distinct cultural practices.

A myth of common descent consists of legends and accounts of communal history that come from the collective experiences of successive generations. It underlies the sense of belonging and similarity for the participants. It provides a rationale and a framework for a growing sense of ethnic identity. Without a myth, an ethnic group cannot define and distinguish itself from other groups and cannot inspire collective action. The second primordial tie is a sense of historical continuity—the construction and reconstruction of a nation's past from an ethnic perspective, which links successive generations. History in this sense tells a story as well as educates the participants. The heroes and the heroines whose deeds it reveals—for example, the British entrepreneur, the Russian worker, and the Chinese peasant—embody the virtues held in high regard by a particular community (Smith, 1987: 24–6). Thus, history enhances the consciousness of ethnicity and deepens the sense of shared identity. A distinctive shared culture, real or imagined, is another bond that helps to bind members of an ethnic group together and to separate them from outsiders. Religion and language are among the common shared cultural phenomena. Other cultural traits, such as customs, institutions, laws, architecture, music, dress, and food, reflect and reinforce the differences. The greater the number of differentiating cultural ties, the more intense the sense of ethnic identity, and the greater the prospects of ethnic persistence (Smith, 1987: 26–8; 1981: 66–8; 1984). Finally, an ethnic community has its attachment to a particular territory or 'homeland' which becomes a symbolic geographical centre and a means of maintaining group cohesion. Members of an ethnic community do not cease to identify themselves with their own group when they have lost their homeland and are scattered around the world, for ethnicity is a matter of myth, memory, and symbol. Some ethnic groups maintain themselves by sustaining a hope for political independence or for the recapturing of a lost homeland.

Ethnic group cohesion produces a strong sense of belonging and solidarity, which in time of threat or outside pressure can override class, faction, or regional divisions within the community. At times, ethnic solidarity may be overlaid by other kinds of loyalty, but from time to time, especially in the face of external enemies, it will re-emerge in sufficient force and depth to override other types of allegiance (Smith, 1987: 30).

Prior to the emergence of the modern nation-state, most ethnic communities lived in autonomous and mainly homogeneous social environments. The clan, the chiefdom, the municipality, and the empire represented such autonomous socio-political units. The empire was the largest of many kinds of political entity, claiming vast reaches of territory and containing peoples of different ethnic origins. Because of its size, it inevitably allowed diverse ethnic groups to remain in essentially autonomous environments. Indeed, those communities situated far from the centre had only vague awareness of imperial law and order. As the empire expanded, it became increasingly difficult to maintain hegemony over its periphery. Thus, the realm of an overextended empire gradually diminished as the cost of coercing allegiance from groups in the periphery exceeded the worth of goods expropriated from their areas. Eventually, the imperial system gave way to the new system of the 'nation-state' (Raymond L. Hall, 1979: xvii–xix; Emerson, 1960; for an account of state formation and pre-existing 'nations', see Smith, 1987: 129–73).[1]

The origins of the transition to nationhood can be traced back to the gradual unification of the territories of England and of France and later to the emergence of unified Spanish, Swedish, and Polish states during the first half of the second millennium AD. In the Muslim world, there also existed some sort of 'centralized states' in Fatimid Egypt, Savafid Iran, and the Ottoman Empire. These unified territories gradually were transformed into 'national states'[2] through the impact of the revolutions (economic, social, and cultural) that occurred in the Western world and through the provision of equal legal rights to various strata of people and the growth of public education systems (Smith, 1987: 138; cf. Kohn, 1967a, 1967b).

The emergence of nation-states in Europe provided a skeleton for the subsequent development of European colonialism around the world. One of the significant results of colonial incursion was that it incorporated diverse peoples, often against their will, into larger political unions. As a consequence, autonomous ethnic communities became, in many cases, 'national minorities'.[3] In order to facilitate control of resources, the colonial state could no longer tolerate ethnic preferences for autonomy. Ethnic autonomy was considered an obstacle to the development of the state (Raymond L. Hall, 1979: xix). Thus, the cultural–political domination of the core ethnic group dictated the forms and content of the social institutions and political life of the whole population within its domain (Smith, 1987: 139) and denied the significance of ethnicity. The new allegiance was no longer to the tribe or religion, but to the 'ideology of the common good', which was based on the idea that diverse communities could be brought together in a new system where the state, rather than ethnicity, was pre-eminent.

However, since ethnic communities reflect group cohesion founded upon traditions and psychological phenomena not shared by others, ethnic behaviour tends to be based on particular interpretations of situations motivated by group self-interest. Members of ethnic communities find themselves divided in their allegiances between loyalty to the state to which they belong and to the communities of their birth and upbringing. This divided loyalty has created fertile ground for separatism, as many ethnic communities struggle to become nations.

The emergence of the new nations of Africa and Asia gave rise to a vast and variegated literature on what has become known as political modernization and development, nation-building, and political integration. Characteristic of these writings is the view that ethnic and primordial sentiments are temporary impediments to the development of modern society, and the assumption that ethnic loyalty will fade away and be replaced by cleavages based on class and occupational status (Weber, 1947; Marx, 1964; Rokkan, 1970). Similarly, there is what sociologists call the 'diffusionist' theory of social integration, associated with such distinguished scholars as Talcott Parsons and Neil Smelser (see, for example, Parsons and Smelser, 1956), according to which the culture and values of the core group in a society are gradually diffused throughout the peripheral areas. For Marxists, too, ethnic and cultural conflicts have always been regarded as a transitional phase of capitalist society, which in the course of events will be overwhelmed by class struggle. Class, according to them must inevitably become the main basis of division between people, replacing such earlier factors as tribe, religion, language, and national origin (Birch, 1978: 325; Glazer and Moynihan, 1975; Nielsen, 1980: 76).

Hence, the persistence of ethnic and nationalistic conflicts has been characterized either as a manifestation of residual loyalties from an early stage of social development, due to dissolve as society becomes more developed, or as essentially a form of economic conflict, though the participants may not recognize it as such. Systems of social organization based on ethnic distinctions are seen to correspond to a relatively primitive level of development and to be poorly adapted to the demands of modernity. They are destined to be replaced by more adequate institutions during the modernization process.

Central to the argument of the modernization literature is the notion that the process of modernization—the spread of a market economy, urbanization, bureaucratization, growth of literacy, and improved social communications—leads to a cultural diffusion that cuts across primordial attachments, producing a more or less homogeneous culture within a given geographical territory.[4] Deutsch (1969, 1966), for instance, maintained that 'When several population clusters are united through more communications or more economic activity, then the people begin to think of themselves as a country'. This implies that the process of modernization in the form of increased social mobilization and change erodes or breaks down major clusters of old social and psychological commitments, replacing pre-existing ethnic bonds with more functional ties (Lipset and Rokkan, 1967), and people become available for new patterns of socialization and behaviour (Deutsch, 1961: 494). For example, the modern industrial society 'teaches people not to take their class position as God-given, but as somehow the result of their own efforts. Class status becomes achieved rather than ascribed or inherited' (Deutsch, 1970: 89). The growing sectors of the national economy attract many of the most gifted and energetic persons from the minority and tribal groups, leaving these traditional communities weaker, more moribund, and easier to govern (Deutsch, 1963: 5). The process of modernization, it was argued, will lead to a transfer of primary allegiance from the ethnic group to the state.

Coleman (1960: 345), Pye (1963: 10), and Weiner (1967) argued that the problems of integration and of creating political consensus in the new states was largely a problem of establishing more effective channels of communication and transportation so that all segments of the society can become more closely involved with each other.[5] A similar assumption was made by Lipset and Rokkan in their study of the relationship between social structures and party system. They wrote:

The National Revolution forced ever-widening circles of the territorial population to choose sides in conflicts over values and cultural identities. The Industrial Revolution also triggered a variety of cultural counter-movements, but in the long run tended to cut across the value communities within the nation and to force the enfranchised citizenry to choose sides in terms of their economic interests, their shares in the increased wealth generated through the spread of the new technologies and the widening markets (1967: 18–19).

Huntington (1971: 289–90) pointed in the same direction when he asserted that modernization is a homogenizing and irreversible process. In his discussions on the 'functional necessity' of the family, Moore (1958: 161–3) suggested that it is possible for the modernized industrial societies of the world to do away with the family and substitute other social arrangements. These functionalist or developmental theories tended to view ethnicity as a threat to the very existence of the state (Olzak, 1985: 73). From this perspective, identification as 'Moro' or 'Malay' clearly competes for the loyalty and commitment with identification as 'Filipino' or 'Thai'.

The Marxist tradition was consistent with these predictions. Its view of the historical evolution of capitalist societies reinforced the view that ethnic divisions were particularly salient in an early phase of economic organization (Marx, 1964, 1969). In a system of communal property, the major element of production was land collectively owned by the community. Tribal or ethnic loyalty was therefore essential. In contrast, the capitalist system emphasized the separation of the elements of production (land, labour, and capital) and thus undermined the significance of community leadership. Capitalism also entails a growing polarization between those who own the means of production and those who do not (Nielsen, 1980: 76-7; see Lijphart, 1977: 54). Thus, class replaces ethnicity as the fundamental social division. In an attempt to synthesize functionalist and Marxist perspectives with respect to collective action, Dahrendorf (1969) and Nielsen (1980: 77) suggested that both the functionalist and Marxist arguments may be roughly summarized as follows: 'Since the criterion of ethnicity is not functionally relevant for the allocation of authority in industrial societies, the likelihood that conflict is based on ethnic boundaries decreases with modernization.'

According to the literature discussed above, the broad process of modernization affects all cultures roughly in the same manner and it is irreversible and unilinear. As a result, all states composed of diverse ethnic groups will eventually become nationally integrated. However, the mounting empirical evidence of ethnic nationalist movements around the world (Esman, 1977) suggests that both the functionalist and the Marxist models are inadequate. Despite a long experience of multi-ethnicity, the integration of diverse social groups into already-established political systems remains unrealized. Many nation-states face challenges by minority nationalist movements of various forms, ranging from the relatively peaceful assertion of separate identity, seeking certain degrees of local autonomy, to violent activities or civil war, demanding political independence. In the second half of the 1970s, for example, there were no less than 117 organized movements for separation and autonomy concentrated in at least 44 nations of the North and South (Boulding, 1979: 276).

Since the middle of the 1960s, the cruder versions of modernization theory have been challenged by many scholars. Connor (1972), for instance, maintained that a global survey illustrated that ethnic consciousness was on the rise as a political force, and that modernization, in the form of increasing social mobilization, appears to have increased ethnic tensions and to be conducive to separatist demands. This is because social mobilization fosters ethnic competition, especially in the competitive modern sector: 'It is the competitor within the modern sphere who feels the insecurities of change most strongly and who seeks the communal shelter of "tribalism"' (Melson and Wolpe, 1970: 1115). Similarly,

Enloe (1973: viii) suggests that ethnic identity fulfils primary and communitarian needs in a society increasingly dominated by complex, large-scale organization. For Geertz (1963: 154), social mobilization 'does not do away with ethnocentrism, it merely modernizes it'. It has even been suggested by Bates (1974: 471) that 'Ethnic groups persist largely because of their capacity to extract goods and services from the modern sector and thereby satisfy the demands of their members for the components of modernity. In so far as they provide these benefits to their members, they are able to gain their support and achieve their loyalty.' It has been argued, further, that ethnic solidarity persists because modernization has been articulated in terms of converging aspirations:

People's aspirations and expectations change as they are mobilized into modernizing economy and polity. They come to want, and to demand, more—more goods, more recognition, more power. Significantly, too, the orientation of the mobilized to a common set of rewards and paths to rewards means, in effect, that many people come to desire precisely the same things. Men enter into conflict not because they are different but because they are essentially the same. It is by making men 'more alike,' in the sense of possessing the same wants, that modernization tends to promote conflict (quoted in Horowitz, 1985: 100).

The following section examines in more detail some of the major theoretical perspectives which argue the persistence of ethnic solidarity in modern societies.

Theoretical Perspectives of Ethnic Solidarity

Two major theoretical perspectives converge upon the assumption that contemporary ethnic solidarity is essentially a product of modernization and is related to and legitimized by nationalism. They diverge in their definitions of ethnicity and in the elements they identify as instrumental in generating ethnic mobilization.

Reactive Ethnicity: Internal Colonialism and the Cultural Division of Labour Theory

Hechter, the chief proponent of the internal colonialism thesis, declares that the great nation-states such as Britain and France not only annexed overseas colonies in the course of their development and expansion but also incorporated diverse 'internal' ethnic communities within their state frontiers (Hechter, 1975, 1978; Casanova, 1963; Stavenhagen, 1965; Dos Santos, 1970). An internal colony is said to exist when the relationships between the core area, which is normally dominated by an ethnic majority, and the peripheral regions, which in many cases comprise culturally diverse minority communities, are characterized by exploitation due to the uneven progress of industrialization and development (Hechter, 1975; Gellner, 1973; Nairn, 1977). The core community attempts to maintain its superior position by institutionalizing the existing inequality. Ethnic solidarity (ethnic mobilization) is seen as a reaction of the culturally distinct periphery against the exploitative core. The theory predicts reactive ethnic solidarity when economic activity within the periphery is concentrated in the hands of the core ethnic majority or when a pattern of structural discrimination exists. Hechter calls this arrangement a 'cultural division of labour', which he defines as a 'system of stratification where objective cultural distinctions are superimposed upon class lines' (1975: 30; see Leifer, 1981). A cultural division of labour occurs 'when individuals are assigned to specific types of occupations and other social roles on the basis of observable cultural traits or markers' (Hechter, 1974: 1154). With cultural division of labour, ethnic boundaries tend to coincide with the boundaries of structural differentiation, and there is a basis for group action along ethnic lines.

Hechter (1973, 1974, 1975) also points out that internal colonialism is one of the main mechanisms leading to a cultural division of labour. Modernization, he argues, does not affect the entire population of the nation evenly, but spreads from one or several centres towards the periphery. As modernization is diffused throughout the nation and the periphery is gradually absorbed into the modern system of production and exchange, the centre controls development in the periphery in such a way that the populations in the periphery are involved in the activities most profitable and complementary to the requirements of the centre and the least rewarding for the periphery. Specialization of the periphery in such activities leads to a pattern of structural discrimination. Whenever the populations of both the centre and the periphery can be differentiated by cultural phenomena, mobilization on the basis of ethnic boundaries is likely to take place. In other words, ethnic solidarity is intensified when cultural and functional lines are concurrent. However, this reactive ethnicity model admits that modernization and functional differentiation in the social system as a whole are not sufficient to explain the persistence of ethnic solidarity in modern societies. Other factors, such as historical development, must also be taken into account.

Competitive Perspectives of Ethnicity

Some social theorists attribute the underlying problem of ethnic conflict to the competition of diverse ethnic groups for scarce resources. In his study of ethnic conflict in Africa and Asia, Horowitz (1971), for example, suggests that a large part of the conflict stemmed from the competition between members of different ethnic groups for positions in the national bureaucracy and in the armed forces. Likewise, Cohen (1969) writes: 'The continual confrontations between the Hausa and Yoruba in competition for economic benefits from trade, together with the rapidly increasing population of Sabo and the complexity of its social life, led to the emergence of a unified "tribal" administration and leadership' (quoted in Dulyakasem, 1981: 15). Lieberson (1970: 9–10), in his study of ethnic relations in Canada, also demonstrates that competition exists between languages in contact because the optimal conditions for their native speakers are usually discordant and because the gains that one language makes are to the detriment of the other.

Like the reactive ethnicity model, the competitive perspectives model provides justifications for the emergence of ethnic mobilization and antagonism in modern nation-states. However, the two models associate such emergence with quite different structural arrangements. For the former, ethnic solidarity emerges when modernization generates a cultural division of labour in which a peripheral ethnic community is in a disadvantageous position and exploited by the centre. The latter relates solidarity to situations in which social mobilization and development spreads more uniformly over the national territory, the distribution of occupations and social roles in the centre and periphery become more equal, and the cultural division of labour breaks down (Nielsen, 1980: 79-80). In other words, when diverse ethnic groups come to compete in the same labour markets and increase their access to similar sets of resources, mobilization along ethnic boundaries is most likely to occur (see Despres, 1975; Young, 1976; Hannan, 1979; Rothschild, 1981; Olzak, 1983). There are two distinct lines of argument within the competitive perspectives: the split labour market and ecological competition theories.

THE SPLIT LABOUR MARKET THEORY

Bonacich (1972, 1976, 1979; Bonacich and Modell, 1980) argues that split labour markets generate ethnic antagonism and ethnic collective action. His central argument is that when the price of labour for the same work differs by ethnic group, a conflict develops which may result in extreme ethnic antagonism encompassing all levels of intergroup conflict, including ideologies, behaviour, and institutions. In split labour markets, conflict develops among three key classes—business (employers), higher-paid labour, and cheaper labour. Employers generally aim at having as cheap a labour force as possible to compete effectively with other employers. If labour costs are too high (for example, due to union action), employers may turn to cheaper sources, importing overseas groups or using indigenous conquered populations. Higher-paid labour is threatened by the introduction of cheap labour into the market. Employers use the cheaper-labour class partly to undermine the position of more expensive labour, through strike-breaking and undercutting. Wilson (1978) has similarly observed that ethnic antagonism is most intense when a labour market is divided along ethnic boundaries: for example, when migrating Blacks from the South of the United States were employed as strike-breakers in the expanding industries in the North, ethnic antagonism resulted. But in contrast to the cultural division of labour model, the split labour market theory stresses class cleavage as the primary determinant of behaviour in a labour market differentiated by wage and ethnic categories.

THE ECOLOGICAL COMPETITION THEORY

Hannan (1979) asserts that ethnic conflict is mainly the result of intense competition between ethnic groups for the occupation of social niches

McCarthy and Zald, 1977; Tilly, 1978; Nagel and Olzak, 1982; ard and Hamilton, 1986). The theory suggests that as the process of economic and political development diffuses more homogeneously over the national territory and breaks down the cultural division of labour, competition between ethnic groups increases (Barth, 1969; Despres, 1975; Banton, 1983). This is because the modernization process favours reorganization within larger boundaries, rather than along kinship, village, or other small-scale lines (Hannan, 1979). The major implication of this theoretical perspective is that, to the extent that an ethnic population competes with the majority or other minority ethnic groups for the same jobs, the resulting niche overlap leads to ethnic competition. Under conditions of stable or shrinking resources, this competition results in ethnic collective action (Olzak, 1985: 76). Meyer and Hannan (1979) and Birch (1978) further argue that because different international economic and political organizations (e.g. the EEC, IMF, NATO, and UN) emphasize membership, negotiation, and power only among nation-states, ethnic mobilization in the post-war period increasingly has reflected demands by sub-national groups for self-determination and sovereignty. Indeed, the ecological competition theory defines the causal relation between modernization and ethnic mobilization both in terms of raising levels of ethnic competition and increasing legitimacy of sub-national and territorial ethnic movements.

In sum, the reactive ethnicity and competition perspective models appear to be useful in explaining the persistence and the emergence of ethnic solidarity (Smelser, 1969; Hilton, 1979; Verdery, 1976) and seem to be applicable to the present situations in the Muslim regions of the southern Philippines and southern Thailand. The more pertinent models, however, are internal colonialism and the cultural division of labour and the ecological competition perspective. The former affirms that the extent of overlap between low cultural status (minority) and low-reward occupations produces ethnic mobilization. Although one may argue that the cultural division of labour in the Philippines and Thailand is not as clearly visible as that in the developed nations, there is no doubt that more Muslims in Mindanao and in Patani (low cultural status groups) perceive themselves to be engaged in low-status occupations compared to the non-Muslim majorities in the two countries. Furthermore, the internal colonialism argument describes a historical process in which ethnic lines and occupational divisions converge (Hechter, 1974, 1975; Nielsen, 1980). Such a scenario seems to conform to the circumstances of the Moro and the Malay communities of the two regions which have been absorbed by the accident of history into the larger Filipino and Thai nations.

The ecological competition perspective model suggests that the penetration by the centre, through economic and political developments, creates the conditions under which ethnicity is regarded as the most viable and effective political instrument for the periphery to oppose the centre. Ethnicity, it is argued, is the most pertinent element of social and political organization where there is rapid economic and political change, where the competition for roles and resources is profound, and where achievements rather than heredity determine access to those roles and resources (Hannan, 1979; Ragin, 1979). The structural alterations resulting from the spread of modernization from the centre lead to ethnic reactions and mobilization. This model (with only slight modification) has already been tested in the four Muslim provinces of southern Thailand and proved workable (Dulyakasem, 1981, 1984).

The split labour market theory is less relevant to the situation in Mindanao and Patani because Muslims in these regions do not receive different wages for the same work.

From the foregoing discussion, it is clear that modernization and change create conditions favourable to a competitive environment and ethnic organization. The question is: under what conditions will an ethnic group choose to take action aimed at separatism as opposed to, for example, working for change within the system? The answer is that it depends largely upon the many specific conditions which catalyse separatist potential (see Anderson *et al.*, 1967; Smith, 1982; McVey, 1984). Ethnic separatism generally refers to ethnic action aimed at complete withdrawal from state-imposed socio-economic and political arrangements. Such action, which is frequently a reaction to assimilation attempts at the centre, aims to promote cultural, linguistic, religious, geographical, and economic autonomy within a specific state or to obtain complete political independence. Separatism is, therefore, a particular form of ethnic conflict which requires specific conditions to generate separatist action. As mentioned earlier, distance from the centre is a first essential condition. Regions on the periphery of the precolonial states in Africa and Asia, for example, could regard separation as a political option whenever a weakening of the centre created the opportunity. Those groups near the centre had no separatist option and could only hope to change the regime at the core (Anderson *et al.*, 1967: 71; Smith, 1982: 32–6). A coincidence between geography and cultural plurality is also crucial. In polities like Malaysia and Lebanon, where the major ethnic groups are scattered throughout the national territory, separation is not a genuine option. The role played by India as midwife to Bangladesh demonstrates that the prospects for successful separatism could be vastly improved by the patronage of a neighbouring country. The breakdown of central institutions also provides an opportunity for separation. Finally, in certain cases, the separatist aspirations of peripheral areas can be achieved simply by not obeying the central authority, but without overt secession. The central government's attempts at coercion are often rendered difficult by terrain and distance (Anderson *et al.*, 1967: 71–4).

By definition, separatism is also a species of nationalism,[6] for it seeks to enhance ethnic autonomy and in some cases to gain political independence and self-determination. Ethnic nationalism often arises in communities with political grievances or discontent resulting from the dominance of ethnic minorities on the periphery by the core group, as in the cases of the Moros in Mindanao and the Malays in Patani. If left unresolved, such grievances are likely to precipitate social disturbances and separatist movements. Local ruling élites resent their loss of status and the correspondingly increased prestige of officials of the central government. Ordinary people resent outside influence on their accustomed way of life and fear that the new ruling groups will disturb existing local interests. Thus, they rally behind their traditional leaders in movements of resistance and opposition to foreign authority. If territory coincides with well-defined boundaries of communal language, religion, or some other ascriptive characteristics, the tendency may well be towards secession or civil war, threatening the very existence of the state (Oberschall, 1973: 44–5).

Among other factors, religion has often played a pivotal role as the revolutionary vehicle for separatist movements (for a discussion of the impact of religion on politics, see Lenski, 1961). In particular, Islam has been an important aspect of many liberation movements, including the Moro and the Malay separatist struggles. As Louis Gerdet observed, 'The nationalist movements among Muslims are essentially Islamic reactions against alien domination' (quoted in Aruri, 1977: 268). In some instances, Islam has become the fundamental ingredient of the struggles; in others, Islamic concepts and symbols are integrated into nationalist dogma to provide movements with the necessary religious coloration to appeal to a wider population. In most cases, Islam and nationalism reinforce each other in their rejection of alien rule.

Islam and Nationalism

Theories of social movement acknowledge that the emergence of an active movement is contingent on the simultaneous existence of several conditions. One such condition is a mobilizing ideology, that is, 'an interconnected set of ideas and beliefs that provide both a call to and a guide for action by defining what is wrong, attributing blame and responsibility, and addressing the ... question of "What is to be done?"' (Snow and Marshall, 1984: 138). The significance of a mobilizing ideology in relation to collective action has long been recognized. As Rude (1964: 74–5) observed with respect to the French Revolution, 'some unifying body of ideas, a common vocabulary of hope and protest' was required to tie together the discontents and aspirations of the people. In the Baltic states, for example, Robert Miller (1986: 57) has noted that 'The most visible sign of revival was the burgeoning of the dissident movement—in some cases primarily religious in motivation, but in all cases centering on demands for national self-expression and resistance to russification.'

In developing nations, religion is often employed as both the mobilizing ideology and the organizational basis for collective action, particularly in response to foreign intrusion and cultural dislocation. That religion can serve as a vehicle to mobilize the masses for revolutionary and nationalist movements is almost self-evident. Cohen (1969: 210) wrote:

Religion provides an ideal 'blueprint' for the development of an informal political organization. It mobilizes many of the most powerful emotions which are associated with the basic problems of human existence and gives legitimacy and stability to political arrangements by representing these as parts of the system of the universe.... Religion also provides frequent and regular meetings in congregations, where in the course of ritual activities, a great deal of informal interaction takes place, information is communicated, and general problems are formulated and discussed. The system of myths and symbols which religion provides is capable of being continuously interpreted and re-interpreted in order to accommodate it to changing economic, political and other social circumstances.

In South-East Asia, for instance, religion played a dominant role in campaigns against the intrusion of different foreign powers. Religious-oriented nationalists and xenophobists fought to defend the old order and the religious values and institutions it supported. The histories of Indonesia, Burma, and Malaysia illustrate this well (see Noer, 1978; von der Mehden, 1968; Geertz, 1960; Sjamsuddin, 1984; Nagata, 1984; Federspiel, 1985). As von der Mehden (1968: 147) observed:

Nationalist leaders in search of an issue with both universal and emotional appeal could find none better than religion. Political capital could be gained by using religious personnel, incorporating religious ideas into the political platform, identifying the party or politician with religion, and capitalizing upon minor and major insults to religion. The nationalists had no monopoly on efforts to employ religion as a tool; it was used also by such diverse groups as colonial administrators and Communist propagandists.

In Marxist literature, on the other hand, religion is viewed primarily as a means of social control and a source of stability; its mobilizing function is, therefore, underplayed (Snow and Marshall, 1984: 139; see Olcott, 1982; cf. Wuthnow, 1980).

In the Muslim world, Islam, as a political ideology, is used for purposes of mass mobilization to achieve varying and sometimes very diverse objectives (Ayoob, 1981: 5). Through its rituals and institutions, Islam mobilizes people for collective action. The mosques, for instance, not only provide a comparatively safe place for people to assemble, but they constitute one of the few places where grievances and revolutionary ideas can be expressed and discussed. Rituals, such as the Friday prayers at the mosques, serve as an important means of mobilizing the masses. Furthermore, the *imam* (prayer leaders) and other religious functionaries provide 'the symbols of integrity and selflessness that contrasted sharply with the corruption and hedonism associated with the centers of temporal power' (*New York Times Magazine*, 11 March 1979); the 80,000 mosques in Iran undoubtedly played a decisive role in mobilization for the Islamic revolution of February 1979.

A review of Muslim history indicates that Islam has often functioned as a mobilizing and radicalizing force in the face of alien domination. For example, Islam made it possible to mobilize millions of disparate Muslims in Indonesia under an organization called Sarekat Islam.

Founded in 1929, Sarekat Islam grew and became the recognized spokesorganization for Indonesian nationalism (von der Mehden, 1968; Noer, 1978). In Algeria, Algerian nationalists utilized religious symbols to mobilize and legitimize their war of liberation against the French (Fanon, 1965). In Egypt, Gamal Abdel Nasser was aware that the Egyptian people could be best mobilized by associating nationalism with Islam. He referred to the Egyptian armed forces as the 'Army of God' and to their confrontations as 'battles of purification' (Haddad, 1980; Lewy, 1974). Even the Soviet system, which appears almost immune to the impact of ethnicity, has been challenged by a rising Muslim ethnic self-assertion (Bennigsen and Lemercier-Quelquejay, 1967; Bennigsen and Broxup, 1983; Lapidus, 1984; Burg, 1984; Wimbush, 1985). Another well-known example of the use of Islam as a mobilizing force for a revolutionary movement was the Mahdist rebellion in Sudan in the late nineteenth century. It was led by a man who claimed to be the Mahdi (saviour prophesied in Muslim tradition) and inspired and legitimized by a concept of *jihad* (see Siddiqui, 1982; Esposito, 1980).[7] The movement involved the Mahdists' opposition to British colonial rule. It succeeded in creating the Mahdist state which, however, was ultimately crushed by the British at the turn of the century. Rejecting the British and their way of life, the Mahdist adherents were urged to 'put aside everything that has the slightest resemblance to the manners of the ... infidels—dress, drum and bugles' (quoted in Snow and Marshall, 1984: 139).

Muslim revolutionary movements of the 1980s, such as those of the Moros in Mindanao and the Malays in Patani, have striking similarities with the movements mentioned above. They are inspired and legitimized in religious terms, and their basic conflict seems to have been cultural rather than economic. As Snow and Marshall (1984: 134) have argued, 'If we examine the rhetoric of revolt, it appears that what is at issue, at least in the mind of the dissidents, is not so much one economic system versus another, but one culture or way of life versus another.' The conflict of culture was also reflected in Ayatollah Khomeini's words when he remarked, 'I will devote the remaining one or two years of my life to reshaping Iran in the image of the Muhammed.... What the nation wants is an Islamic republic, not a democratic republic, not a democratic Islamic republic. Just an Islamic republic ...' (*Time*, 12 March 1979; see Khomeini, 1981; Algar, 1983). Given the fact that Islam does not clearly separate religion and politics (Jansen, 1979; Laffin, 1979), it is only natural that Islam, as 'a total and unified way of life', plays its role in all community activities, including revolutionary activity.

What does Islam teach Muslims with respect to revolutionary activity? Islam is a religion as well as a socio-political order—*deen wa dawla*. Ever since the establishment of the first Muslim community in Medina, Islam has always motivated Muslims to engage in political struggles, liberation movements, and wars. Its primary concern is that Muslims must strive for the betterment of life by dealing practically and comprehensively with it in all its essential aspects. Muslims are urged to strive, particularly when they are in an underprivileged position. Those who are fighting for freedom or self-determination are regarded as striving in the cause of Islam. To borrow Bowles' words, the cause of freedom is the cause of God.

Islam regards freedom and liberty as the fundamental right of every man. According to Khalifa Abdul Hakim (1974: 220), the Preamble of the American Declaration of Independence essentially repeats the Islamic thesis. The Quran repeats the story of Pharaoh and Moses to emphasize that a great prophet is always a liberator. Whenever the weak and the helpless are persecuted, the honest and righteous must rise to restore social justice. There is no place in Islam for the dictatorship of any individual or group.

Additionally, Islam views some Western political ideas and codes of law as inconsistent with the *sharia* (Islamic law). Thus, a struggle is often called to preserve the glory of Islam by re-establishing a community based on the *sharia*. This struggle is considered by Muslims to be a divine commandment that cannot be disobeyed (Cudsi and Dessouki, 1981: 5-11). As Khomeini (1981: 126) declared, 'It is our duty to work toward the establishment of an Islamic government.'

Another indirect dimension of Islam in relation to revolutionary action is that it serves as a means of defining identity and linking the Muslim community with a larger Muslim world. For example, the Islamization of the Moros in Mindanao and of the Malays in Patani created a break with older cultures and brought an intellectual and rational impetus that set in motion the process of revolutionizing their world view (Al-Attas, 1969: 5). Islam supplanted many cultural practices, such as eating and clothing habits and ceremonial observances. One of the significant cultural phenomena directly effected by the influence of Islam in both regions was the use of the Jawi script, and liturgical Arabic entered their languages (cf. Yegar, 1979). This led to the emergence of a new sense of ethnic identity in each region that helped to distinguish the Moros and the Malays from the Filipinos and the Thais respectively.

Also, Islamization resulted in an ideological bond among the Muslims, creating a sense of community that transcended local and provincial loyalties. It established linkages with outside Muslim political communities that facilitated the Islamic struggles (Suhrke and Noble, 1977: 182). Jamaluddin Al-Afghani (1839-97) characterized Islam as the greatest bond which linked the Turk with the Arab, the Persian with the Indian, and the Egyptian with the Maghrebite in opposition to European intrusions (quoted in Aruri, 1977: 267). In this sense, the Moro and the Malay communities are part of the 200 million Muslims of the Malay race in South-East Asia. Muslims of the two major Malay nations, Indonesia and Malaysia, have been concerned with the situation of their 'unredeemed brethren' under Filipino and Thai rule (cf. Pitsuwan, 1982: 258-9). These Malays, at the same time, belong to the wider world of Islam. If Moros met Berbers at the Kaaba in Mecca, each not knowing a word of the other's language and therefore incapable of communicating orally, they would nevertheless regard each other as 'brother', because of the sacred Kaaba they shared.

Finally, Islam has been a unifying force and, in the face of imperialistic expansion, has provided the spirit for revolutionary action. In Islam, there is 'a moral bond, a vital cultural identity that would help rejuvenate the patriotic zeal needed for nationalist struggle against the imperialists' (Bayat, 1980: 99). Aruri (1977: 267) has also pointed out that Islam provides a basis upon which Muslims come together in an effort to create a viable political existence and to confront foreign rule.

1. A 'nation' refers to a relatively large group of people, with a consciously shared set of cultural patterns, symbols, and other expressions of solidarity, inhabiting a given territory and obeying the same laws and government (cf. Rejai and Enloe, 1969: 143; Linz, 1973: 37; Patterson, 1977: 67–9; Anderson, 1983: 15). A 'state' is defined as a bureaucratic administrative unit, which extends over geographical territory, mobilizes forces of violence and possesses legitimate authority to use such forces (Weber, 1947). Some theorists assume that 'nation' also implies collective identity. To the extent that collective identity overlaps with bureaucratic legal boundaries, a 'nation-state' is said to exist (Olzak, 1985: 72).

2. Smith (1987) employed the term 'national state' to distinguish it from 'nation-state', as the latter implies congruence and co-extensiveness between the territorial state and the ethnic population and culture. In this book, however, the two terms are used interchangeably.

3. According to Rose and Rose (1972: 2), the term 'national minorities' was employed first in Europe to describe the social position of some groups of people in relation to the rest of the population. In many countries, people of certain cultural backgrounds had ancient attachment to a given place, and the land they occupied bore their name. However, some groups of people found themselves within political boundaries in which the majority group was a different nationality; that is, the territory of a state did not coincide with the territory inhabited by the historical nationality groups (cf. Krejci and Velimsky, 1981: 32–42).

4. 'Modernization' comprises a number of related processes of social and economic change, including industrialization, urbanization, increases in literacy and the provision of education, expansion of the mass media, modifications of traditional relationships, and rising expectations. Some of these have also been referred to as 'social mobilization' (Milne, 1981: 83). Eminent social scientists, such as Karl Deutsch and Samuel Huntington, have been concerned with the political accompaniments of modernization; some of the consequences of this are discussed here.

5. The complex tasks of integration involve 'the development of a sense of nationality by subsuming various cultural loyalties; the integration of political units into a common territorial framework; the greater linkage of the rulers and the ruled; the integration of citizens into a common political process; and, finally, the working together of individuals for common goals' (Weiner, 1967; see Jahan, 1972).

6. Nationalism can be defined as 'an ideological movement for the attainment and maintenance of autonomy, cohesion and individuality for a social group deemed by some of its members to constitute an actual or potential nation' (Smith, 1976: 1). It is both an ideology and a movement that aspires to nationhood for the chosen entity (see Smith, 1979, 1982; Kautsky, 1971).

7. The term *jihad*, which is commonly translated as 'holy war', literally means 'exertion'. Its juridical–theological meaning is exertion of one's power in Allah and making His word supreme throughout this world. There are four different ways in which a Muslim may fulfil his *jihad* duty: by his heart; his tongue; his hand; and by the sword. It is the fourth meaning that inspires armed fighting or war (Khadduri, 1964: 168–9).

1
Moro and Malay Societies: Historical, Economic, and Social Backgrounds

The Mindanao and Patani Scenes

THE incorporation of the Moros[1] of Mindanao and the Malays[2] of Patani into the Philippine and Thai states respectively, was the end result of centuries of struggle. It was accidents of political history which placed the hitherto autonomous Muslim communities under alien rule.

The basic concern of the Muslims, ever since they were conquered, has been to ensure their survival as a distinct Muslim *ummah* (community), a community committed to its conception of the ideals of Islam. As Gowing (1975: 28) has pointed out: 'The Muslim minorities in the Philippines and Thailand ... are caught in the dilemma of having to reconcile the demands of their rather traditionalist conception of faithfulness to Islam with the demands of citizenship in modern states in which non-Muslims predominate.' However, the ever-tightening control that the central governments in Manila and Bangkok have been exerting over the areas has made the Muslims' goal of regaining their autonomous status more difficult, if not impossible, to attain. In fact, many Muslims feel that the governments have gradually threatened their survival as a distinct Muslim community. This, together with other factors, has contributed to ethnic tension between Muslims and non-Muslims which intermittently has erupted in local uprisings. The tension also has its roots in the Moro and Malay histories and in socio-economic and cultural characteristics that serve to perpetuate a sense of separate identity from the mainstream of the Filipino and Thai populations. Geographical proximity, cultural affinities, and seasonal migration of labourers have helped to keep alive a strong identification with neighbouring Muslims of Malaysia and Indonesia. Moreover, they all belong to the same Shafiite sect of Sunni Islam, the prevalent sect of South-East Asian Muslims.

An examination of some of these distinct characteristics of Muslims in Mindanao and Patani might, therefore, provide insights into the present Muslim separatist problems.

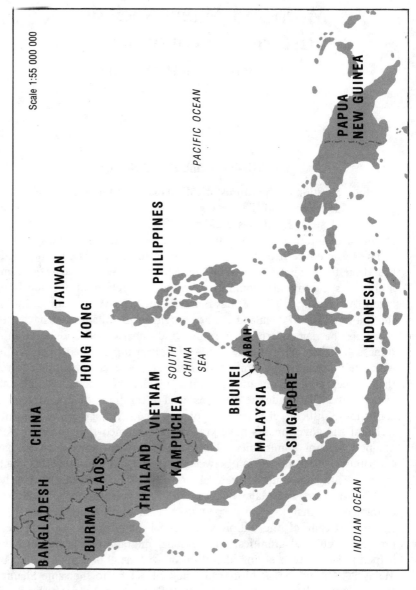

Map 1 South-East Asia

Mindanao: History, Economy, and Society

The History

Mindanao is a variant form of the name 'Maguindanao', which connotes 'inundation by river, lake and sea'. The region comprises the principal island of Mindanao and a chain of some 369 smaller islands of the Sulu Archipelago. With an area of about 96 438 sq. km—about 33 per cent of the total land territory of the Philippines—it contains 22 of the nation's 73 provinces.

The Islamized groups of the Indonesian–Malay race (Hooton, 1947: 639–42), who are popularly known as 'Moros', were once the majority inhabitants of the Mindanao region. However, as a result of the colonization of Mindanao, the socio-demographic status of the Moros has changed from majority to minority. Table 1.1 illustrates the estimated size of the present-day Bangsa Moro cultural–linguistic groups in Mindanao.

The 13 cultural–linguistic groups of Muslims are concentrated in western and southern Mindanao island, the Sulu Archipelgo, and coastal areas of southern Palawan. Only five (Sulu, Tawi-Tawi, Lanao del Sur, Maguindanao, and Basilan) of the twenty-two provinces have a Muslim

TABLE 1.1
Population of Bangsa Moro Cultural–Linguistic Groups in Mindanao, 1980 and 1982

Group	Population Size (1980)	Percentage of Mindanao Population	Estimated Population Size by Moro Respondents (1982)
Maranao (Malanao)	742,962	6.81	986,197
Maguindanao	644,548	5.91	1,328,908
Tausug (Joloanos)	502,918	4.61	1,527,891
Samal (Sama)	244,160	2.24	656,827
Yakan	196,000	1.80	124,039
Sangil (Sangir)	77,000	0.71	79,595
Badjao (Samal Laut)	28,536	0.26	22,932
Kolibugan (Kalibugan)	15,417	0.14	23,954
Jama Mapun (Samal Cagayan)	14,347	0.13	94,305
Ilanun (Iranun)	12,542	0.12	436,932
Palawanon (Palawani or Pinalawan)	10,500	0.10	10,813
Kalagan	7,902	0.07	8,907
Molbog (Melebuganun)	7,500	0.06	9,658
Bangsa Moro population	2,504,332	22.96	5,310,958
Total Mindanao population	10,905,243		

Sources: Philippines, National Economic and Development Authority (1980a); Abbahil (1983).

Map 2 The Philippines

majority. In order to appreciate these changes, it is necessary to discuss, in broad outline, the history of the Moro region.

Although most literature—both Muslim and non-Muslim—on pre-Islamic Mindanao is ambiguous, the Moros place great importance on the fact that their traditional literature, *tarsila*, attests to their having established rights of domain in the Mindanao region far back in the pre-Islamic era. Before the coming of Islam, for example, Jolo and Cotabato had already become important trading centres. More importantly, the Muslims consider themselves as having entirely separate origins from the Christianized Filipinos (Ruth L. P. Moore, 1981: 36–40), despite the fact that they are of the same race.

Accounts of the spread of Islam in Mindanao seem no less speculative than those of the pre-Islamic period. Nevertheless, it is evident that Islam came to the Philippine Archipelago well before the arrival in 1542 of Villalobos and his reconnoitring expedition force, who named the islands 'the Philippines' in honour of Philip, the son of Emperor Charles V and heir to the Spanish throne (D. G. E. Hall, 1981: 272). Arab merchants and later Muslim preachers, who had settled along the trade routes between the Red Sea and China, were responsible for the spread of Islam in South-East Asia. It is likely that commercial relations between the Muslim empire and the Malay world, including the Philippine islands, were established during the expansion of Islam from the eighth to the tenth centuries.

Chinese sources dating from the Yuan Dynasty (1280–1368) record trade activities between China and Sulu. Jolo was a commercial centre to which Arab, Thai, Indonesian, and Indian traders also came. It is suggested that Arab traders had established their settlements in the Mindanao region at the end of the thirteenth century (Majul, 1973: 63–4). By 1450, a Muslim sultanate had been founded among the Tausugs at Buansa (Jolo) by a Mecca-born Arab trader, Syed Abu Bakr (Saleeby, 1913: 10–11). Under the direction of Sultan Syed Abu Bakr, the study of Islam was begun, political institutions were shaped along Islamic lines, and preachers were sent out from Buansa to convert the people of the surrounding areas. At its height, the influence of the Sulu Sultanate extended from Basilan and the Sulu Archipelago to Palawan, the coast of southern Zamboanga, and Borneo (May, 1981: 213).

Fifty years later, the Maguindanao Sultanate was established around the mouth of the Pulangi River on Mindanao island. Sharif Muhammad Kabungsuwan, a Muslim preacher of Arab–Malay parentage from Johore, is credited with its foundation. From there Islam spread along the coast to the Gulf of Davao and inland to Lake Lanao and Bukidnon. Though no specific date is known for the Islamization of the people of Lake Lanao, the Maranao *tarsila* trace their Islamic legitimation as well as royal lineage back to the same Sharif Muhammad Kabungsuwan. In essence, the Moroland of the mid-sixteenth century was in the process of becoming part of the wider Muslim world of South-East Asia. Commercial relations and political alliances linked the Moro sultanates of the Mindanao–Sulu region with other neighbouring Muslim states.

The Islamization of Mindanao and Sulu resulted in an ideological bond among different groups of people in the region which led to the emergence of a new sense of ethnic identity that distinguished Muslim from non-Muslim populations. Islam also became a unifying force and, in the face of colonial domination, provided the basis for revolutionary action (Al-Attas, 1969: 8; cf. Majul, 1974: 6). It was this underlying ideology of resistance that conceived a *jihad* against Spanish colonialists when they arrived in the Philippine islands with their sword and Cross in 1565.

The objectives of the Spanish colonization of the Philippines were twofold: to increase the territorial domains of the Crown and to Christianize the local population. These objectives brought the Spaniards into conflict with the Muslims, who had established political and commercial dominance and were spreading Islam in Mindanao. Moreover, the bitter experience of the long Iberian crusade (*c.*711–1492) against the Moors led the Spanish to regard the Muslims of this region, whom they called 'Moros', as enemies. In the central and northern parts of the Philippine islands, Islam had had very little impact upon the indigenous inhabitants, who were mostly animists and nature worshippers (D. G. E. Hall, 1981: 272–3), and the Spaniards found it relatively easy to convert them to Christianity. The Christianized natives were called 'Indios'.

In pursuing their goals, Spanish forces dislodged Muslim communities in Manila, Mindoro, and the Visayas during the first decade and a half of their arrival. They also dispatched a series of military expeditions against the Moros in the Mindanao–Sulu region. Since Islam had already taken a firm hold in the Moro area, the Spanish imperialistic policy only served to strengthen Muslim resistance and to provoke raids of reprisal against the Spaniards. Consequently, the Spaniards and the Moros were in an almost continuous state of battle, raid and counter-raid for more than 300 years (Majul, 1973). Despite a number of military successes, especially after the introduction of steam-powered gunboats in 1848, the Spanish forces were not able to win the so-called 'Moro Wars'. Thus, Spain never actually achieved sovereignty over Moroland. Nevertheless, she did succeed in restraining the spread of Islam through the Philippine Archipelago.

Three aspects of the Moro Wars proved to have enduring consequences. First, the Spanish deliberately fostered religious antagonism and a derogatory image of the Muslims in order to mobilize the Indios to fight wars against the Moros. This created a bitter enmity between the two groups, despite the many similarities in their socio-cultural heritage. By the end of the Spanish period, a dichotomy had developed which was regarded by both sides as inextricable. Indeed, the Moros were viewed by the Christian Filipinos as a 'separate *rasa* (race), with all that term's connotations of elementality and primordiality' (quoted in Bentley, 1982: 62). Secondly, the earlier social relations between Moros and pagan tribes who later became 'Indios' were strained, though the Muslims had never managed to conquer them (Schlegel, 1979a: 2).

Thirdly, beginning in the second quarter of the nineteenth century, Spanish authorities had begun a strategy that entailed the large-scale relocation of Christian Filipinos from the overcrowded and poorer islands in the north to the sparsely populated frontier of Mindanao in order to colonize it by 'proxy'. By the end of the Spanish period, there had been substantial Christian migration to Moroland.

Following the defeat of Spain in the Spanish–American War, the United States took over the Philippines as successor to Spain under the Treaty of Paris, signed in 1898. Included in the Spanish cession was Moroland, even though much of its territory had never been incorporated into the Spanish colony. As will be argued in more detail in Chapter 2, the change of colonial master brought no major improvement in the Moro position. The American government, like the Spanish, adopted a policy aimed primarily at incorporating Mindanao into a wider Philippine state. The process of integrating the Moros and the Christians of the Philippines, however, was not seen by the Americans as a religious matter. The United States repeatedly presented its policy as one of greater tolerance of Moro religion and customs. It saw its major programmes toward the Moros as entirely secular: 'American laws and notions of justice were imposed; public schools and other services were created; land was surveyed and titled; and homesteaders were encouraged to migrate from overpopulated northern areas and settle on open land in Mindanao' (Schlegel, 1979b: 99).

From the Moro point of view, these integration programmes were a threat to the survival of the Muslim community, and they resisted them with the same tenacity and the same sense of religious obligation with which they had resisted the Spanish policies. It took the American military authority in what became Moro Province (1903–13) a whole decade of warfare to 'pacify' the Moros. The subsequent civilian administration of the renamed Department of Mindanao and Sulu introduced a 'policy of attraction' under which reforms were implemented in the Moro region.

The Commonwealth Government (1935–46) saw Moroland as a land of promise and opportunity. It replaced the policy of attraction with one of political and economic integration aimed at integrating Moroland into a future independent Philippines. Many Moros were opposed to the policy and wanted no part of the Christian Filipino nation (Jubair, 1984: 65). Consequently, a new wave of anti-government activity erupted, and thousands of Muslims died resisting the incorporation of Mindanao into the Philippine Republic. None the less, when the Philippines was granted independence in 1946, Mindanao was included. American interests were seen to be best served in an undivided territory, sacrificing Moro interests and aspirations to live in a separate sovereign country.

As a consequence of decades of integration effort both during the American colonial period and under the Philippine government, some Moros have been 'integrated' into a larger Philippine national identity, while the great majority of them, particularly those in the rural areas, have no real sense of being part of the Filipino nation. While the Christian Filipinos look to the West and to Christian communities, Moro roots are in Islamic civilization whence they draw their values, laws, and sense of history. These roots have grown deeper since the resurgence of Islam after the Second World War. Many new mosques and *madaris* (religious schools) were built in Moroland; hundreds of young Moros left Mindanao to study in the Middle East; Muslim preachers from Egypt and other Muslim countries increased in numbers; and Muslim societies, such as Ansar El-Islam, the Muslim Association of the Philippines, and the Sulu Islamic Congress, were organized. Meanwhile, the process of expropriation and marginalization of the Muslim population reached unprecedented intensity, especially after President Ferdinand Marcos came to power. By the end of the 1960s, the hostility and distrust which for centuries had characterized relations between the ruling Christian Filipinos and the ruled Moros intensified. The resistance of the Moro people to these nefarious designs on their homeland had also achieved new momentum. This led to a full-fledged struggle for the separation of Mindanao in the 1970s. Another important factor behind the rising tension between the two groups was the economic exploitation of Mindanao.

The Economy

The colonization of Mindanao has contributed to a steady deterioration in the economic conditions of the Moros. The long war with Spain resulted not only in the reduction of the Moro population but also in the disruption of its economy. Spanish attempts to disrupt trade between the Moro sultanates and the outside world had undermined their relative prosperity. The situation further deteriorated during the American period as foreign interests sought to exploit the vast agricultural, mineral, and forest resources of Mindanao. Resettlement projects and restructuring of land rights were introduced in the region as part of this design.

From 1912, agricultural 'colonies' (Christian settlements) were established in different parts of Moroland. The Moro and other indigenous peoples suffered from these arrangements, particularly through the enactment of laws pertaining to land acquisition in Mindanao. The Public Land Act of 1919, for instance, declared that all lands within Philippine sovereignty were public domain and that ownership was a state-granted privilege. This meant that all Moro ancestral landholdings, which had been passed down from generation to generation as *pusaka* (inherited property), were no longer valid. The government reserved to itself the power to issue titles to public land, possession of which became proof of ownership. Under this Act, a Christian Filipino could apply for up to 24 hectares of land while a non-Christian was limited to only 10. Most Muslims, with the exception of some members of the Muslim élite, failed to secure a title to their ancestral lands. Whereas prior to 1912 the Moros owned most of the land in Mindanao and Sulu,

TABLE 1.2
Estimated Moro and Non-Moro Populations in Mindanao, 1903–1980

Year	Mindanao Population	Moro Population		Non-Moro Population	
		Number	As Percentage of Mindanao Population	Number	As Percentage of Mindanao Population
1903	327,741	250,000	76	77,741	24
1913	518,698	324,816	63	193,882	37
1918	723,655	358,968	50	364,687	50
1939	2,244,421	755,189	34	1,489,232	66
1948	2,943,324	933,101	32	2,010,223	68
1960	5,686,027	1,321,060	23	4,364,967	77
1970	7,963,932	1,669,708	21	6,294,224	79
1975	9,146,995	1,798,911	20	7,348,084	80
1980	10,905,243	2,504,332	23	8,400,911	77

Sources: Philippines, National Economic and Development Authority (1980a); Gowing (1977, 1979); Abbahil (1983).

in 1972 only 30 per cent had land in their name. By 1982, the percentage of Moro landowners had been reduced to 17 (Jubair, 1984: 54–7).

Some of the Christian migrants settled in the present-day provinces of Lanao del Sur, Maguindanao, Sultan Kudarat, and North Cotabato, displacing numerous Muslims in the process. This process intensified during the post-colonial era. By 1975 there were at least 37 colonies of Christian settlers, comprising 50,247 families and involving 695 500 hectares of land (Silva, 1979: 46–7). In addition to these government-sponsored settlers, millions of Christians migrated on their own initiative. Through these resettlement programmes and land policies, the demographic composition of Moroland has changed so that Muslims have been reduced from a majority to a minority. Table 1.2 compares the size of the Moro and non-Moro populations in Mindanao between 1903 and 1980.

Whereas migration and related land laws were seen by the Muslims as 'legalized land-grabbing', designed by the colonialists for the sole benefit of the Christian Filipinos, the American and the Philippine administrations saw them as serving several purposes. First, they offered a solution to the emerging problems of the overpopulated northern islands. Secondly, it was argued that the Moros and other tribal groups in Mindanao would benefit from the example of the technologically advanced and more industrious Christian migrants. Thirdly, the exploitation of the natural resources of Mindanao would add to national wealth and to the profits of American investors. Fourthly, the colonies of Christians would tie Mindanao more closely to Manila, providing some counterweight to the Moro link with neighbouring Muslim nations. Finally, it was believed that through sheer submergence of the Moro population, the problem of Moro dissidence would be solved (May, 1981: 216).

Along with these migrants, American and, to a lesser extent, Japanese

TABLE 1.3
Utilization of Cropland in Mindanao, 1977 (Export Crops)

Crop	Area under Cultivation (ha)	Percentage of Mindanao Cropland
Coconut	1 403 896	40.3
Sugar	58 516	1.7
Rubber	49 392	1.4
Pineapple	37 302	1.1
Abaca	64 384	1.9
Coffee	60 594	1.7
Banana	27 458	0.8
Fruits	10 843	0.3
Others	39 244	1.1
Total	1 751 629	50.3
Total Mindanao cropland	3 485 277	

Source: AFRIM Resource Center (1980).

corporations penetrated Mindanao soon after free trade between the United States and the Philippines was instituted in 1909. The main industries which began to operate during the first few decades of the American occupation were pineapple, rubber, and coconut processing. Other significant industries which emerged after the Philippines gained independence included logging, bananas, sugar, coffee, mining, and fishing. These industries then, as they still are now, were based on the needs of the advanced industrial countries. The production of goods for the export market became the major concern of the economy, and large export volumes were seen as a measure of development. This has been the pattern of the Mindanao economy. Tables 1.3 and 1.4 show the utilization of cropland and Table 1.5 the export products of Mindanao.

As Tables 1.3 and 1.5 indicate, 50 per cent of the cropland in Mindanao in 1977 was taken up by export crops; about 25 per cent of the

TABLE 1.4
Utilization of Cropland in Mindanao, 1977 (Local Consumption)

Crop	Area under Cultivation (ha)	Percentage of Mindanao Cropland
Rice	747 274	21.4
Corn	744 410	21.4
Banana (non-export)	117 529	3.4
Root crops	80 686	2.3
Vegetables	32 906	0.9
Fruits	10 843	0.3
Total	1 733 648	49.7
Total Mindanao Cropland	3 485 277	

Source: AFRIM Resource Center (1980).

TABLE 1.5
Mindanao Exports, 1977

Product	Value (US$)	Percentage of Mindanao Exports	Percentage of National Exports
Coconut products	337,203,250	43.3	10.7
Banana and pineapple	144,680,000	18.6	4.6
Wood products	143,190,500	18.4	4.5
Metallic minerals	131,670,000	16.9	4.2
Others	22,702,840	2.9	0.7
Total	779,446,590	100.0	24.7

Source: AFRIM Resource Center (1980).
Note: Percentages of Mindanao exports do not add up to 100.0 because of rounding.

total value of exports from the Philippines came from Mindanao. The questions are not whether Mindanao should produce for export, but rather who benefits from export earnings and whether domestic resources are being allocated so that the needs of the local population are met. *Mindanao Report* issued in 1980 by the AFRIM Resource Center in Davao City pointed out that Mindanao's economic development, which is being stimulated by international corporations, spreads its benefits unevenly. It benefits a small number of individuals who comprise the élite. Its principal concern is to generate a peripheral type of capitalist growth, which will be dependent on the advanced economies of the industrial nations for investment, technical assistance, and markets. In agriculture as in industry, the large, medium, and small businessmen depend in one way or another on linkages with foreign capital in pursuing their activities. Even companies which are 100 per cent Filipino-owned rely on foreign loans to survive. The incentives being offered to attract such investments include political stability, cheap wages, an emasculated labour force, and generally lower costs of production. The overall effect of externally imposed economic development has resulted in increased polarization between the privileged few and the disadvantaged masses (AFRIM Resource Center, 1980; see David, 1977).

This pattern of economic development disrupted the Moro economic order, by replacing subsistence production with export-oriented production, and drove the Moros to the economic periphery. The shift from subsistence agriculture to industrial or agricultural wage employment should have been accompanied by a rise in the standard of living for the population in general. But it did not occur for the Moros. First, circumstances made it both unnecessary and unfeasible from the point of view of foreign corporations in Mindanao to use the Moro people as their primary labour pool. The influx of Christians from the north provided an ample labour force for the industries. The colonial entrepreneurs already had a long-standing relationship with the Christians of the north and thus preferred them to the Moros (Rocamora,

TABLE 1.6
Average Gross Regional Domestic Product and Per Capita Income
in the Muslim Area (Regions IX and XII), in Mindanao,
and in the Philippines (peso)

	Muslim Area (Regions IX and XII)	Mindanao	Philippines
Per capita income (1975)	799	803	1,308 (est.)
Family income (1975)	5,343[1]	5,199	5,840
Per capita GRDP (1977)	959	1,240	1,769 (est.)

Sources: AFRIM Resource Center (1980); Philippines, National Economic and Development Authority (1979, 1980b, 1983a).
[1]The slightly higher average family income in the Muslim area (Regions IX and XII) may be attributed in part to the development of the logging and wood products industries, agricultural plantations, and the industrial growth of Iligan and Zamboanga Cities, most of which are non-Muslim undertakings.

1981: 245-6). Most plantation and company workers are Christians; in such industrial centres as Davao City, Iligan City, and Cagayan de Oro, the Moros make up only a very small proportion of the labour force. Secondly, since the industrial corporations in Mindanao are export-oriented, their products are not made for local consumption. Consequently, the industries are not concerned with the level of income of the local population. Table 1.6 compares Gross Regional Domestic Product (GRDP) and per capita income in the Muslim areas (Regions IX and XII),[3] Mindanao, and the nation.

Most Moros still adhere to traditional means of livelihood such as farming, fishing, and small trade. Tenants and smallholders often fall prey to money-lenders who make cash advances at exorbitant interest rates. The *prenda* (mortgage) system of usury usually results in smallholders losing their lands to the usurer. Some Moros own land without the benefit of a title. Through government policies and laws, the Moros have been steadily dispossessed of their holdings. Most of the wealth in Moroland is in the possession not of the Muslims but of the Christian Filipinos and foreign entrepreneurs. The influx of Christian settlers into the Muslim region since the 1950s has reduced the size of Moro landholdings to below subsistence level in most communities (Ahmad Domocao Alonto, 1979: 59). The Muslims are mostly rural-dwellers in a subsistence economy. Moro traders find it difficult to compete with the more enterprising businessmen from Luzon and the Visayas. Moro fishermen have been threatened by fishing companies which operate in the Sulu Sea, the Moro Gulf, and Basilan Strait. Foreign investment in the fishing industry has increased over the years, especially since the proclamation of the Fisheries Decree in 1975, which allows foreigners to invest in fishing. Although there are Moros in the government sector, most government offices are dominated by Christians. This state of affairs is compounded by the low educational profile of the Muslims. In the field of modern education, the Moros are a generation behind the Christians (Tongson, 1973: 20). This is mainly because of the lack of proper educational policies suited to the temperament and demands of the people. As Glang (1969: 81) wrote, 'How can we encourage the Moro children to appreciate school education, to love their Christian brothers when all that they read and learn in their books shamefully points to their ancestors and forefathers as pirates, bandits and cutthroats?' The Moros are perceived as 'good for nothing'; 'the only good Moro is a dead one'. No less than 250,000 Moros have already migrated to the Malaysian state of Sabah (Komite ng Sambayanang, 1981: 271), and more than 80 per cent of the male migrants have gone to seek employment (Abduljim J. Hassan, 1978: 113) and to escape from the violence at home.

It should be noted, however, that the process of Moroland's administrative and structural integration into the Philippine state provided the opportunity for some Moro leaders, particularly the traditional élite, to retain their past prerogatives and position of dominance and become part of the national élite through national and local elections and appointments to high government positions. They allied themselves with Manila and were regarded by many separatists as 'collaborators' with an 'oppressive regime' and as part of the problem in Moroland (Mastura, 1980: 112; Gowing, 1979: 57).

In the view of many government officials, it was inevitable that Moroland would be at a disadvantage in certain aspects. While the Indios submitted to the colonial masters and were laying the foundations for economic growth, the Moros were devoting their resources to fighting against colonialism. Nevertheless, they stress that the government is trying hard to uplift the economic condition of the Moro community: at least 31 presidential measures have been passed, and millions of pesos have been allocated for development projects in Mindanao–Sulu (see Chapter 5). The government's effort, however, has had limited success. Very little of what is given benefits the Moros directly, and some of the funds appropriated for projects in Moro areas, such as roads and schools, find their way into the pockets of dishonest local politicians (Gowing, 1979: 185). Moreover, the unpreparedness of the Muslims for change, particularly the detachment of lands from the traditional system of community and clan ownership and the fear of displacement and domination by the non-Muslims, have caused them to resist Manila's efforts. Indeed, one commentator has been led to conclude that the Moro community remains so economically stagnant that '... many believed that the Moro people were more developed during the Spanish days than at present' (Jubair, 1984: 65).

The Society

Islam has been an important component in the 'chemistry' of Moro society. It introduced many of its features to the Moro people of various cultural groups and integrated them into a single society distinct from those non-Muslims in the Archipelago. Before the advent of Islam, for example, the Moros were divided among numerous *timuway* (headmen). Islam transformed these *timuway* into the *datu* (men of rank), and the powerful *datu* of large *barangay* (communities) developed into the sultans of Moro society.

At the time of Spanish contact, Moro society consisted of three social classes: aristocrats, commoners or freemen, and slaves. Ideally, aristocrats were the descendants of Syed Abu Bakr and Sharif Muhammad Kabungsuwan, the first sultans of Sulu and Maguindanao respectively. Aristocratic status was hereditary, though some gained status by their own personal skills and attributes. In this category, sultans and *datu* were the two dominant ranking aristocrats. The lesser noble in the aristocratic class was called *dumatu*. Second in social standing were the commoners who constituted the major portion of the Moro population—farmers, fishermen, sailors, and artisans. The indentured freeman or debt-bondsman (*ulipun*) was considered the lowest rank in the Maguindanao system. Commoners were also followers (*sakop*) of sultans and *datu* and owed their loyalty and service to them (Gowing, 1979: 52). In return, they received protection and shelter from their sultans and *datu* when needed. Moreover, commoners enjoyed a certain measure of social mobility. For instance, a successful *nakuda* (leader of trading expedition) could obtain enough wealth and followers to improve his social status. The lowest class was the slaves (*baniaga*). They were mainly obtained by taking captives in wars or raids. Slaves were regarded as the property of their masters. Male slaves normally worked in the fields while females performed household tasks and occasionally became the concubines of their masters. Since Islam does not encourage slavery, it was possible for slaves to earn their freedom by arrangement with their masters. Former slaves and their descendants blended into general society and eventually became commoners (for a comprehensive discussion on Moro socio-political organization, see Stewart, 1977: 276–306; Beckett, 1982; Warren, 1982; Saber, 1976). Kiefer (1972a: 23–30) has suggested that in the middle of the nineteenth century, the population of the Sulu Sultanate was approximately 400,000, of whom about 1 or 2 per cent belonged to the aristocracy; 10 per cent were slaves; and the rest were commoners (*tau way bangsa*).

Mednick (1957: 43–6) characterized the Moro political structure as a pyramidally arranged hierarchy of authority. At the top of the hierarchy was the Sultan or *Raja* (ruler), the head of state from whom all authority was derived. In each political unit (district), there was a *panglima* (sultan's personal representative) serving as chief administrator. Assisting him in different administrative duties were 'a *nakib* who was in charge of military matters, a *laksamanna* who was the *panglima*'s messenger, a *perkasa* who was his aide-de-camp, a *bintala* who was the district supervisor of priests, and so forth' (Mednick, 1957: 44). The Sultan also formed the *Rumah Bechara* (House of Discussion) comprising a group of powerful aristocrats or *datu*. Members of the *Rumah Bechara*'s council performed different duties and held various positions, ranging from that of Prime Minister to that of Inspector of Weights and Measures.

Paralleling the political hierarchy was the structure of religious authority. At the top of the hierarchy was also the Sultan. He thus functioned as the head of both religious and political authorities. Under the Sultan was a *qadi* (judge) who was his religious supervisor and the chief authority in matters of Islamic law (*sharia*). The *qadi* was one of the most powerful authorities in the Moro society. Under certain circumstances, for example, a *qadi* could override decisions of the Sultan (Mednick, 1957). At district level, the *pendita*, a man well versed in the Quran and the Prophetic traditions, served as religious adviser to the *panglima*, the district chief. There were also religious functionaries who commanded considerable influence in the society. They were the *imam* of different community levels. The most influential *imam* was the one who was appointed to the Sultan's mosque, and the least influential ones were those of the local village mosques. Each *imam* was assisted by a *khatib* (preacher) and *bilal* (prayer caller).[4] The *imam*, *khatib*, and *bilal* are the principal personnel of the mosque.

It should be reiterated that although the Moros constituted a single society and nationality, they comprised different ethno-cultural groups, as indicated in Table 1.1. The Maranaos, Maguindanaos, and Tausugs were the major groups, constituting 75 per cent of the total Moro population. The development in social and political structure of the three was similar but not identical. The position of *panglima* in the Sultanate of Sulu, for instance, carried certain prerogatives and powers which differed somewhat from those exercised by the *panglima* in the Sultanate of Maguindanao (Gowing, 1979: 45). It is believed that the institution of the sultanate and the formal offices under it were more developed in Sulu than in Mindanao (Majul, 1964: 3). Not all Moro groups had sultans and the number of sultanates varied. For example, the Tausugs had only 1 sultanate; the Maguindanaos had 3 major sultanates; more than 40 sultanates were identified among the Maranaos, 15 of which were considered pre-eminent (Saber, 1974, 1976). The Moro sultanates were also characterized by a kinship system which polarized loyalties and interest along blood lines. It caused rivalries and dissension. As discussed in Chapter 2, these loosely knit sultanates were not always united during their struggle against colonial powers.

In contemporary society, Moros have managed to retain some of their traditional socio-political structure. The datuship, for instance, has remained a powerful institution and has been able to adapt itself to the political machinery of the modern Philippines (discussion of the contemporary Moro society with emphasis on leadership is provided in Chapter 4).

It is important to note that the foundation upon which Moro society was established is Islam. All laws, with the exception of those which concern some traditional customs, were essentially within the bounds of the *sharia*. And most authorities, from the sultan to village headman, were persons of religious inclination. Thus, Moro society was governed by Islamic and customary laws and religiously inclined leaders. Because of this strong religious inclination, the Moros have been less ready to absorb and adopt more modern ways of life, even after their inclusion into the Philippine nation. Many Moros regard the structure of the Filipino government, its codes of laws and political ideas, as inconsistent with the *sharia*.

Patani: History, Economy, and Society

The History

Although the Malay-Muslims comprise only about 3 per cent of Thailand's 50 million predominantly Buddhist population, they constitute a large majority in the four southern provinces of Patani,[5] Narathiwat, Yala, and Satun, previously known as 'Patani Raya' or 'Greater Patani'.

Patani was historically the ancient Malay kingdom of Langkasuka (Chinese: Lang-ya-hsiu). Founded sometime in the first century AD, Langkasuka was considered one of the important commercial ports for Asian mariners, especially those who sailed directly across the Gulf of Siam from the southern tip of Vietnam to the Malay peninsula (Teeuw and Wyatt, 1970: 1–2). When the major powers in the region, such as the Cambodian empire of Angkor, the Mon empire of Pagan, and Srivijaya, intervened in the peninsula's affairs, the small states struggled to keep their independence; some maintained a precarious existence while others were submerged. In the case of Langkasuka, the kingdom gradually disappeared and was replaced by the kingdom of Patani.

The *Hikayat Morong Mahawangsa* (*Kedah Annals*) attributes Patani's foundation to people from Kedah, while *Phongsawadan Muang Patani* (*Chronicles of Patani*) gives credit to a ruler from an unidentified Kota Mahligai. In *Sejarah Kerajaan Malayu Patani* (*History of the Malay Kingdom of Patani*), Ibrahim Shukri suggests that the ruler of an inland city of Kota Mahligai developed a small coastal village into a bustling port at the expense of the former. Subsequently, Kota Mahligai was abandoned in favour of the new city, which was Patani. The date of this event is not certain. Tome Pires, a Portuguese who passed through Malacca in 1511, wrote of Patani as though it had been in existence before 1370. On the other hand, the region was still identified by Chinese sailors as Langkasuka at the time of the voyage of Admiral Cheng-ho around the year 1403.

Patani has been held to be one of the cradles of Islam in South-East Asia. The initial contact of Patani with Islam was undoubtedly a by-product of Arab trade with China. Arab and Indian merchants had settled in the commercial centres of Patani by the end of the twelfth century, intermarrying with the indigenous people and forming the nucleus of a Muslim community. More than three centuries after Islam had spread into the area, the Court of Patani was converted to Islam. The kingdom of Patani, it is believed, was officially declared an Islamic

Map 3 Thailand

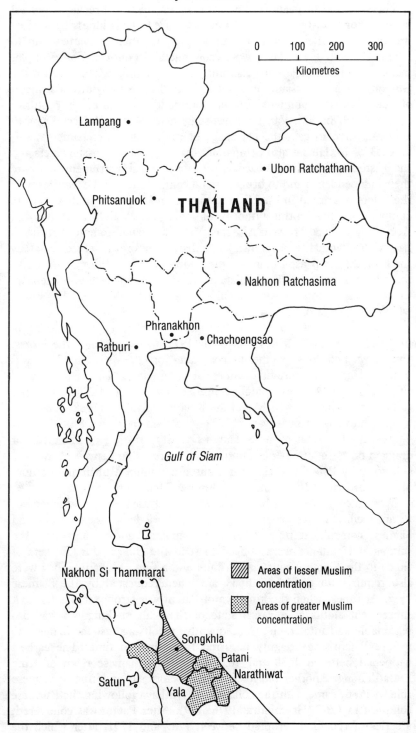

state in 1457 (Teeuw and Wyatt, 1970: 75; Ibrahim Shukri, n.d.: 34).

The adoption of Islam by the Court of Patani was a result of several factors. For one, the King of Patani was cured of his illness by a Muslim preacher, Sheikh Said or Safi-uddin, who then converted him to Islam. Other likely motives were economic and political. For example, Islam provided a means for claiming certain trading privileges with the Gujarati, Arab, Persian, and Turkish traders who controlled much of the Western commerce in the fourteenth and fifteenth centuries (Weekes, 1978: 253). Van Leur has suggested that the Malacca Dynasty adopted Islam to gain the patronage of the Muslim merchants (Majul, 1962: 375-6). He has also pointed out that some of the coastal chiefs and the aristocratic families utilized Islam as a political instrument to secure their independence and to bring about a confirmation of the legitimacy of the rulers interested in establishing their own kingdoms as they were in conflict with the central authority of Majapahit (Majul, 1962: 376-7). According to Syed Hussein Al-Attas, Van Leur failed to note that 'Islam has an "individuality of its own" and that, therefore, it served to satisfy some of the deep aspirations of the people' (Majul, 1962: 381). Thus, the adoption of Islam by the Patani courtiers might have also been because of aspirations by the Patani folk for new values and life-styles, which the new faith offered.

The Muslim kingdom of Patani grew both in population and in prosperity. 'Patani', writes Newbold (1971: 67), 'was once the largest and most populous of the Malay states on the peninsula.' It was described as an important commercial centre for Asian as well as for European traders. The kingdom comprised two major Muslim dynasties: the Patani Dynasty (?-1688) and the Kelantan Dynasty (1688-1729).

The outbreak of civil war in 1729 set Patani in chaos; Sultan Alung Yunus, the last ruler of the Kelantan Dynasty, was killed. The civil war dragged on for a long period, until an aristocrat from Dawai, Mayo, was able to unify Patani and establish himself as Sultan Muhammad. Sultan Muhammad ruled Patani until it fell under Thai control in 1786.

Ever since the establishment of Muslim dynasties, the kingdom of Patani seems to have experienced alternate periods of independence and Siamese control. At times when it was under Siamese suzerainty, the sultans and sultanas were obliged to send the *Bunga Mas* (Flowers of Gold) to the Siamese court as a tribute and a sign of loyalty. They were also required to provide military aid when requested by the Siamese king. At times, when the kingdom of Patani was strong, it was felt that subservient status as a vassal state of the Siamese kingdom was unreasonable and intolerable. Consequently, revolts against the Siamese to restore Patani's sovereignty occurred from time to time. The earliest happened between 1630 and 1633 during the Siamese reign of King Prasat Thong. Another bid for freedom, before surrendering to Siamese rule in 1786, came with the collapse of Ayuthaya following the Burmese conquest in 1767 (Haemindra, 1976: 199). After Patani was conquered, a series of rebellions erupted between 1789 and 1791, after which the ruler of Patani, Tengku Lamidin, was captured. Dato Pengkalan, a respected Malay leader, was appointed by the Siamese authorities as Tengku Lamidin's successor, but in 1808 he also rebelled.

Faced with recurring rebellions, Siam decided to reduce Patani's strength by applying a policy of divide and rule. In 1816 the Patani region was divided into seven provinces: Patani, Nhongchik, Raman, Ra-ngae, Saiburi, Yala, and Yaring. However, the policy of divide and rule did not promote political stability and failed to bring about effective control over revenue collection and taxation of the area. Most of the *raja* found themselves dissatisfied and distressed under the increasing control of Bangkok. Encouraged by disruptions in Kedah, then also a vassal state of Siam,[6] the indigenous Patani rulers revolted against Siam in 1832. They were defeated; nevertheless, there was yet another abortive revolt in the region six years later.

In time, however, the territorial division of Patani resulted in its weakening and becoming more dependent on Bangkok. A reversal of the policy of sending Thai officials to rule Patani proved to be effective in reducing disorder in the region. Thereafter, Bangkok was able to deal with the provinces on an individual basis and a peaceful atmosphere prevailed for several years.

The peace was broken when King Chulalongkorn (1868-1910) introduced the policy of administrative centralization known as the *thesaphiban* system, which aimed at exerting greater direct control from Bangkok. In 1901 Siam regrouped the seven provinces of the Patani area into one administrative unit called 'Area of the Seven Provinces' (*boriween chet huamuang*) and placed it under the control of an Area Commissioner (*khaluang boriween*) who was responsible directly to the Ministry of the Interior. Tribute was no longer required but the treasuries of the Malay provinces had to be handled by the Revenue Department as in other Siamese provinces. By 1906 the area of the seven Malay provinces was administratively reorganized into a 'circle' (*monthon*) called Monthon Patani. The Monthon Patani incorporated the seven provinces into four larger provinces: Patani, Bangnara, Saiburi, and Yala.[7] A district of Kedah was also incorporated into Siamese territory and named Satun Province. Kedah, on the other hand, was ceded to England together with Perlis, Kelantan, and Trengganu in accordance with the Anglo-Siamese Treaty signed in March 1909.[8]

This administrative reorganization of the region ultimately forced the local Malay *raja* and the indigenous ruling élites to acknowledge Siamese authority. When Great Britain recognized Siamese sovereignty over the area in 1909, the former kingdom of Patani was completely dominated by Siam in the same way as other states in Asia, Africa, and Latin America were colonized by the major powers in the nineteenth and twentieth centuries.

Because the Malay-Muslims differ from the Thai-Buddhists in many respects—ethnically and culturally—and because they were forced to become citizens of a state with which they could not identify, they

TABLE 1.7
Gross Provincial Product and Per Capita Income in the Patani Region,
in the Southern Region, and in Thailand, 1980

	Patani Region (4 provinces)	Southern Region (14 provinces)	Nation-wide (73 provinces)
GPP (million US$)	152.28	253.56	406.06
Per capita (US$)	481.41	616.96	619.78

Source: Compiled from information obtained from Office of National Socio-Economic Development Commission, Thailand, in 1980.

resented Thai rule. As efforts were made by the Thai government to assimilate them, resentment developed into political resistance and separatist movement with sporadic insurrections and a series of abortive revolts.

The Economy

In spite of its relatively small size, the Patani region is rich in natural resources. It consists of many fertile plains and valleys and fishing grounds, both along the coasts washed by the South China Sea in the east and the Andaman Sea in the west. The mineral deposits include tin, gold, wolfram, manganese, and natural gas. Despite these natural resources, the economy of the Malay-Muslims has been in a very unsatisfactory condition, compared to the rest of the nation. According to M. Ladd Thomas (1975: 7–8), the real income of the Muslim villagers has decreased precipitously since the mid-1950s. Table 1.7 shows the differences in average Gross Provincial Product (GPP) and per capita income between the Patani area, the Southern region, and the nation.

The majority of the Muslims in the four provinces in the Patani region live in villages along the coasts, scattered through the alluvial rice lands, and, more sparsely, in the rugged rubber-growing interior. Table 1.8 indicates the distribution of the region's population.

TABLE 1.8
Area and Population of the Muslim Provinces of Southern Thailand

Province	Area (sq. mi.)	Population Total	Muslim (per cent)	Rural (per cent)
Patani	812	467,621	77	91
Narathiwat	1,799	469,735	78	87
Yala	1,821	291,166	63	76
Satun	1,076	179,565	66	89
Total	5,508	1,408,087	71 (Average)	86 (Average)

Sources: Thailand, National Statistical Office (1982); Siamban (1985).

Most of the Malay-Muslims are self-employed. Many own smallholdings on which they frequently plant rubber, coconut, and other tropical crops such as rambutan and durian, while others grow rice or are fishermen.

Rice cultivation was formerly one of the main occupations of the Malay villagers, who produced enough rice for their own family consumption. This situation is, however, no longer true. Growing population pressure on cultivable land has resulted in a steady decrease in the size of the average rural landholding (M. Ladd Thomas, 1975: 7–8; 1982: 157–8). In addition, rice farming in the Muslim provinces is almost entirely at the mercy of the climate. Irrigation systems remain inadequate, though they have been introduced by the government. The Malay farmers, therefore, do not produce enough rice to meet local needs. The majority of Muslim villagers purchase their daily rice requirements from the Chinese or Thai 'middlemen'.

Rubber and coconut are two major plantation crops in the region. However, large rubber and coconut plantations are often owned by Chinese and Thais. Most Malays own smallholdings. These smallholders cannot withstand price depression due to middleman manipulation and the fluctuation of the international market. Moreover, some Muslim rubber plantation owners have not been able to participate in government rubber replanting programmes designed to raise the level of production because of 'the lack of trust and communication between the government officials and the Malay people' (Pitsuwan, 1982: 20).

Fishing is another important occupation of the Muslims in the coastal provinces of Patani, Narathiwat, and Satun. A considerable volume of fish from the area is exported to Malaysia and neighbouring provinces each year. With the introduction of a large-scale fishing industry in the region, however, this means the livelihood of the Muslim villagers is threatened. While the Malay fishermen still use small fishing boats (*perahu kolek*) and traditional means of fishing, the Chinese and Thais employ trawlers equipped with more modern technology. The Malay fishermen also depend on the Chinese businessmen to determine the price of their catch.

These occupations of the Malay-Muslims, furthermore, are seasonal in nature. For example, rubber tapping can be engaged in for only about 6 months in a year. In rice farming, the farmers actually work in their paddy fields for not more than 3 months; fishermen can go out to the sea for less than 7 months. Hence, most Muslim villagers live at a subsistence level of existence and find themselves in a very difficult economic situation. Some respondents estimate that about 200,000 Muslims from this region presently seek temporary employment in Malaysia and Singapore. Some have left the provinces in pursuit of employment opportunities in Saudi Arabia and other Arab countries (information obtained from interviews with several Muslim villagers, teachers, and government officials in Narathiwat, May 1985).

On the other hand, the Chinese, who constitute less than 6 per cent of the total population in the four provinces, own and operate most of the business concerns (M. Ladd Thomas, 1975: 7). Most of the Thais are government officials, but there are also some merchants and plantation owners. The Thais and the Chinese own most of the tin mines and the large rubber and coconut plantations, while most of the Malays are village fishermen, rubber smallholders and tappers, small-scale farmers, small shopkeepers, vendors, and labourers. Very few have been admitted to the Thai civil service, and even fewer own and operate large-scale businesses or plantations (M. Ladd Thomas, 1975).

In the 1970s and 1980s, the central government has recognized that increasing separatist activities as well as banditry in the Muslim area are closely linked to the general deterioration of the economic condition there. Several government programmes designed to improve socio-economic conditions in the provinces have been initiated. Under some programmes, for example, rubber plantation owners are given incentives to replace the old trees with a higher-yielding variety. The rearing of livestock has been encouraged extensively. Roads, schools, and other public utilities have been constructed (see Chapter 5). An irrigation system as well as projects to control flooding have been started. In an effort to improve rural life, Mobile Development Units under the guidance of a Community Development Officer (*patthanakorn*) are sent out to the rural areas to institute development projects (Haemindra, 1977: 103). All these programmes are aimed at enhancing the Muslim minority population's loyalty to and empathy towards the Thai government.

However, certain programmes are viewed by the Malay-Muslims as having negative effects on them. For instance, the Sarit administration (1957–63) initiated a project to redress the population imbalance between the Thai-Buddhists and the Malay-Muslims in the Patani region. Under this project, called *nikhom sangton-eng* (Self-help Colony), the government provides 18–25 *rai* (7–10 acres) of land to each family of qualified settlers. Since 1961, some 100,000 Thai-Buddhists from different parts of the country have been relocated to the four provinces. The Thai government's target is to transfer at least 650,000 Thais into the Muslim area (*Journal of Muslim World League*, April 1981; see Che Man, 1985: 103). The project is seen by many Muslims, especially the separatists, as a scheme to submerge the Muslim population and to make the Patani region a predominantly Buddhist area.

Be that as it may, the efforts by the central government to remedy the deteriorating economic conditions in the area have so far met with limited success. Reasons for this include inefficient and irresponsible local government officials; the inadequate funds provided; and perhaps the complex nature of the economic problems. The failure has left many Malay-Muslims with the impression that they are not only being ignored, but are being exploited by the Bangkok government. As Pitsuwan (1982) suggests, 'Since the early 1960's when the central government embarked upon a full-fledged national development program, more efforts have been directed towards the construction of infrastructure to harvest and exploit the natural riches of the area.'

The Society

The indigenous 'pre-modern' society of early Langkasuka was gradually transformed into a more complex one when it was exposed to the major religions of 'highly developed' cultures, such as Hinduism, Buddhism, and Islam. The Islamization of Patani supplanted much of the existing Hindu–Buddhist culture and shaped the area into the *tanah Melayu* (the lands of the Malays) in the peninsula. Thus, the nature of the sociopolitical structure of Patani was typical of Malay society.

The Patani community, like the Malay communities in the peninsula, was generally divided into two basic social classes: the rulers and the ruled. The following general description of the Malay society by Gullick (1958: 66) applies also to Patani. The apex of the ruling class was the sultan, the head of state and the point of reference by which, directly or indirectly, the members of the ruling class determined their relative status. The ruling class comprised persons of both royal and non-royal descent. The royal members of the ruling class were aristocrats who were related to the sultan by blood and were able to aspire to the throne or lesser royal offices. They bore an honorific to signify different royal ranks. The following honorifics, in descending order, were recognized in Patani society: *Tengku* or *Tuan*, *Nik*, and *Wan*. If a woman of royal descent married a commoner, her child no longer bore the royal title, and the marriage was regarded as a 'social degradation' to the woman. But the royal title could be transmitted to children of a male of royal descent by his non-royal wife.

The non-royal members of the ruling class were religious leaders and functionaries and officials of high rank who were absorbed into the ruling class because of their ability and merit. Some of them carried titles such as *Dato' Seri*, *Dato'* and *To'* to symbolize their authority. Sheik Said, the Muslim who converted the ruler of Patani to Islam, for example, was given the title of *Dato' Seri Raja Faqih*.

The subject class was basically divided into three categories of people: subjects (*rakyat*), debt-bondsmen (*orang berhutang*), and slaves (*hamba*). The subjects were men of common birth of all occupations. They were also followers of the sultan and the aristocrats of high rank. The system was based on mutual need. While a subject needed a strong benefactor, an aristocrat needed on his side a private army (Gullick, 1958: 104). The relationship between subjects and rulers in Malay society is best demonstrated by the following expression—although in this case not from Patani: 'I was one of the Mahraja Lela's followers. I must do what he bids me. I am his *ra'ayat* (subject). I would not dare to resist him' (quoted in Gullick, 1958: 65). The subjects were led and controlled by the ruling class. It was difficult for them to rise into the ruling class; only those of exceptional ability, such as religious learned men (*orang alim*) and successful traders (*orang kaya* or *nakuda*), were absorbed into the ruling élite.

The second category of subject class was debt-bondsmen. A debtor who could not repay his debt on time was brought to the house of the creditor (usually members of the ruling class) and became a debt-bondsman. He was bound to take any order and do any work his creditor might demand, until his debt was paid. Since his work and services did not reduce his debt, the right of his redemption existed only in theory (Gullick, 1958: 99). The institution of debt-bondage, therefore, was used by members of the ruling class as a means of obtaining followers to increase power and prestige.

The lowest group of subject class comprised slaves. There was no difference in the nature of the work which the slaves and the debt-bondsmen did for their masters, and the former could under certain circumstances regain their freedom by arrangement or by some act of grace on the part of their masters. Since Islam prohibits the enslavement of Muslims by other Muslims, slaves in Patani society were mainly aborigines (*orang sakai*) and non-Muslim captives of wars.

At the height of its power, the Patani political system, like other Malay states in the peninsula, was influenced by Islam and the earlier kingdoms of western Indonesia. At the top of the hierarchy was the sultan, the head of government in state and religious matters. In each territorial district (*daerah*) there was a prince or a chief who represented the sultan as local ruler. His consent was required for important decisions such as the declaration of war and the signing of treaties (Bonney, 1971: 9–10). At the state (*negeri*) level, there was a hierarchy of officials who ran the government under the direction of the sultan and princes of the royal house. Among the principal officers were the following. The *bendahara* (prime minister) was the executive head of government. The *temenggong* (minister of war) was responsible for maintaining order and functioned as the chief of police and the master of ceremonies at the royal court. The *laksamana* (admiral) was the commander of the warships. The *bendahari* (treasurer) controlled all revenue and looked after the sultan's palaces and equipment. The *shahbandar* (harbour master) was entrusted with the management of the markets and warehouse and of keeping check on the weights, measures, and coinage in use (Gullick, 1958: 7–8; 1981: 11–18; Bastin and Winks, 1979: 16–21). Below them were different titled nobles whose positions stemmed from privileged association with the royal family. A meeting of these officials and nobles constituted a council of *Majlis Bechara* (Assembly of Discussion) in which decisions were made by consensus (Andaya and Andaya, 1982: 47).

There was also a hierarchy of religious authority. As ecclesiastical head, the sultan had a *mufti* (chief judge) as principal religious counsellor. The *mufti* was the highest state official in religious *fatwa* (religious ruling) and Quranic interpretation. *Fatwa* issued by *mufti* could override decisions of the sultan if such decisions were incompatible with the *sharia*. Under the *mufti* was a *qadi*, a man well versed in Islamic law. The *qadi* functioned as Islamic judge of the district and as religious adviser to the district head. The *imam*, *khatib*, and *bilal* of the various mosques were religious functionaries of different degrees of influence in

TABLE 1.9
Provincial Administrative Units of the Muslim Provinces of
Southern Thailand

Province	Number of Districts	Number of Sub-districts	Number of Communes	Number of Villages	Number of Dwellings
Patani	8	3	111	570	71,622
Narathiwat	10	2	70	431	81,092
Yala	6	–	53	281	47,363
Satun	4	2	34	239	26,329
Total	28	7	268	1,521	226,406

Sources: Thailand, National Statistical Office (1982); Siamban (1985).

the community. The most influential *imam* was that of the state mosque. In addition, Patani of the nineteenth and early twentieth centuries was dominated by *pondok* (traditional Islamic schools)[9] which provided Muslim villagers with religious education, scholarship, and moral guidance. Almost all aspects of community life revolved around *pondok* and mosque. *Tok guru* (traditional religious teachers) and *ustaz* (non-traditional religious teachers) who taught in *pondok* were among the most influential leaders in Patani society (cf. Winzeler, 1974: 265–8). The Patani sultanate, like that of the Moros, was a religious-oriented society. It was founded upon the *sharia* and the existing pre-Islamic customary practices (*adat*) and was governed by religiously inclined Muslims.

Following the incorporation of the former Patani sultanate into the Thai nation-state at the beginning of this century, a step was taken by the Bangkok government to bring the Muslim area into the national system. Table 1.9 illustrates the territorial administrative units of the Patani region.

Because the Malay-Muslims had no experience with the Thai system of government, it was necessary—so it was argued—to place them under a Thai-Buddhist bureaucracy. The traditional leaders of the former Patani sultanate were removed and by-passed in the establishment of the national political structure in the region. To this day, very few of the government bureaucrats are Malay-Muslims, although it has long been an expressed policy of the central government to recruit more Muslim officials to deal with the four provinces (Suthasasna, 1984: 237). For instance, no Malay-Muslim has been appointed Governor of the four Muslim provinces, with the exception of Governor Termsak Semantarat who was appointed as Governor of Patani for a short period of time after the Muslim disturbance against the Thai government in Patani in 1975. Of 28 district and 7 sub-district officers in the four provinces, only 2 and 1 respectively are Malay-Muslims. And there are two Muslim mayors from the total of seven municipalities in the region. To some Muslims, the Bangkok government has imposed upon them a system of governmental administration which represents only the interests of Thai-Buddhists. Along with administrative integration, there have been government efforts at cultural and economic integration to counter the centrifugal forces of regionalism and ethnic nationalism. On the one hand, the Muslims have been under constant pressure from different government integration policies and programmes. On the other, they are isolated from the mainstream of Thai social mobility and have become underprivileged minorities. Their subordinate status extends to every sphere of life, including politics, economics, and culture.

That isolation is also reinforced by the Muslims' social and cultural situation. The majority of the Malay-Muslims are rural-dwellers in *kampung* (villages). Most of the villagers' affairs—ceremonies, feasts, and economic pursuits—assume religious significance and involve only people of the same faith. The Muslim social circles are, therefore, confined within their own community. Broadening their social web would mean associating themselves with the Thai-Buddhists, whom they consider as incompatible as 'oil and water'. Intermarriage between members of the two societies is, therefore, uncommon.

The relationship between the Malays and the Thais, especially between villagers and government officials, often produces unpleasant experiences. This is rooted in the cultural discrimination which the Thais have developed with regard to the Muslims and in the nature of the Islamic consciousness of the Muslims (Gowing, 1975: 31). The Malays are popularly referred to as *khaek* (alien or visitor), which they resent, for they are in fact neither aliens nor visitors. They have inhabited the region they occupy for centuries. Officially, the Malays are called 'Thai-Muslims', a deliberate effort to blur their true identity—Malay-Muslims. The equally unflattering attitude from the Muslim side is that the Thai-Buddhists are considered as *kuffar al-harb* (non-believers of the first category), against whom war, according to Islamic law, is permissible. The Thai bureaucrats are sometimes referred to by the Muslims as *to'na* (master) which implies 'colonizer'.

The language barrier is another cultural characteristic that causes the two sides to remain segregated. Despite the fact that the Thais have dominated the bureaucracy in the Malay provinces for more than eight decades, very few bureaucrats understand and speak the local language. Likewise, the Malay villagers are too proud of their own cultural heritage to learn Thai and to accept it as their language. Consequently, official contacts and meetings between Thai officials and the Malay villagers require mediation and 'linkage'. The government's main 'linkages' are the *kamnan* (commune headman) and *phuyaiban* (village headman), the two lowest officials in the government administrative hierarchy. Most of the 268 *kamnan* and over 1,000 *phuyaiban* in the four provinces are Malay-Muslims whose selection has to be approved by a district officer of the Thai bureaucracy. These Muslim *kamnan* and *phuyaiban* fail to create strong links between the government and the villagers, due mainly to the general distrust of the Malay villagers towards the government officials. With such weak 'linkages', the gap between the two separate communities remains; their mutual misunderstandings and prejudices continue.

In contrast, Thai-Buddhists and Chinese (either Thai citizens or resident aliens) have become socially integrated. These non-Muslims, who constitute about 28 per cent of the region's population, are mostly found in towns where the latter primarily engage in business, from shopkeepers to owners of bus companies, and the former serve as civil servants, from schoolteachers to governors. In some provincial and district towns, such as Yala, Satun, and Betong, they form the majority. One or two Chinese or Thai families are sometimes found living in a village with the Muslim villagers. They usually own smallholdings or some of the fishing boats; they may also function as fish merchants, buying the catch from the Malay fishermen and selling it in the towns. Considerable social interaction takes place between these urban-based Chinese and Thai-Buddhist communities, in part because of the compatible nature of their customs and religious beliefs but mainly because the Chinese, unlike the Malays, are willing to be integrated into Thai society. Almost all of the Chinese have learned to speak Thai and have Thai names, and intermarriage between the two population groups is very common.

The Mindanao and Patani Scenes Compared

The Islamization of, and the establishment of Muslim sultanates in, Mindanao and Patani can be regarded as one of the significant historical events of the respective regions. It led to the emergence of a new sense of ethnic identity that distinguished Muslim from non-Muslim communities. It became the root cause of the Moro and the Malay resistance movements against foreign domination (see Chapter 2) and later of separatist struggles against Filipino and Thai rule (see Chapter 3). The two regions, however, differ in some aspects. While Mindanao consisted of a number of ethno-cultural groups and sultanates, Patani had only one sultanate within a single ethnic group. Perhaps, this can be explained by the fact that the spread of Islam in Mindanao was contained by the Spanish colonial campaigns and thus it was unable to develop to its fullest extent in order to completely transcend the local and provincial identity of different Moro groups. The existence of many Moro sultanates and different ethno-cultural groups caused divisions within the Moro resistance movements; it might be argued, however, that such divisions have been a source of weakness as well as strength. Perhaps the Moros were able to resist Spanish campaigns for more than three centuries precisely because they were not united under one single group. In contrast, the Patani sultanate, despite having no ethnic division, suffered many defeats during its resistance campaigns against the Thai kingdom.

The economies of Mindanao and Patani have been dominated by non-Muslims since the beginning of the twentieth century when the regions were formally incorporated into the Philippine and Thai states respectively. Much of the wealth in Mindanao belongs either to Christian Filipinos or foreign investors. Export-oriented development has not elevated the position of the Moro population. Similarly, most of the wealth in Patani is in the hands of Chinese and Thais. Unlike Mindanao, 'land-grabbing' in Patani is not very obvious because the extent of migration and resettlement has not been as great. In addition, there are few multinational corporations operating in Patani. Nevertheless, most Moros and Malays remain in the subsistence economy and adhere to traditional means of livelihood.

The traditional Muslim societies in Mindanao and Patani were similar in many aspects, though they were not identical. The two had basically similar political and social structures, which were essentially founded on Islamic principles and the existing pre-Islamic Malay traditions and customs. The most significant contrast between the two societies is that the Moros' political institution of datuship remains strong in modern Philippine society, whereas the religious institution of *pondok* and religious leadership is still influential among the Malays in Patani. As mentioned earlier, the Spanish colonial campaigns contained the development of Islam in Mindanao, while allowing the pre-Islamic political institutions to flourish. On the other hand, Islam in Patani was able to grow because Siam's main concern, unlike that of Spain, was not religious conversion but political domination. When Patani was last conquered by Siam in 1786, the Siamese attempted to weaken the Patani; in 1816 the Siamese succeeded in dividing it into seven smaller ones, and in 1909, when Patani was formally incorporated into the Siamese state, those Malay institutions were replaced by the Siamese bureaucracy.

A central theme of this study is that the Muslim communities in the Philippines and in Thailand make use of their history, culture, and social conditions, with myths and symbols, to consolidate their struggles, to ensure the persistence of their *dar al-Islam* (territory of Islam), and to develop and choose the course of their opposition to government integration efforts and to the overall political atmosphere of their respective countries. In order to fully appreciate their endeavours, this book will first compare the history of their resistance movements.

1. The term 'Moros' in this book refers to Muslims of various cultural–linguistic groups in the Mindanao–Sulu region of the southern Philippines. It is derived from the ancient Mauri or Mauretania and was sometimes used to denote the Muslim conquerors of Spain. Among some Spaniards, in the absence of a more accurate term, the term was loosely used to refer to any Muslim. The Moros generally referred to themselves as 'Muslims'.

2. The term 'Malays' applies to Patani Muslims of southern Thailand who are Malays, while the term 'Thai-Muslims' means non-Malay Muslims or Muslims who consider themselves Thais. The Muslims in Thailand are divided into two categories: the Malays and the non-Malays. The Malays form the majority, while the Thais, Pakistanis, Indians, Chinese, and others constitute roughly 40 per cent of the Muslim population.

3. The Moro population is concentrated in Western Mindanao (Region IX) and Central Mindanao (Region XII) which consist of ten provinces. In 1977 five of these provinces had predominantly Muslim populations. Sulu and Lanao del Sur were 99 per cent Muslim; Tawi-Tawi had 97 per cent; Maguindanao 86 per cent; and Basilan 67 per cent. Other

provinces with large Muslim populations were Lanao del Norte with 36 per cent and North Cotabato with 17 per cent (AFRIM Resource Center, 1980: 6).

4. The office of *bilal* was named after Bilal bin Ribah, an Ethiopian black slave who became the first prayer caller or *muazzin*.

5. 'Patani' is the Malay version; 'Pattani' is transliterated from the Thai spelling. In this book, 'Patani' is employed.

6. Thailand was called Siam until 1939 and also from 1946 to 1949. In 1939, and again in 1949, it was changed to Thailand, a name which carries a certain note of Thai-people nationalism and irredentism.

7. Bangnara Province was officially renamed Narathiwat Province on 10 June 1942. Saiburi Province was dissolved and incorporated into Patani Province on 16 February 1931.

8. The present Malaysia–Thailand border was established by an Anglo-Siamese Treaty of 1909 which resulted in the transfer of suzerainty over Kedah, Kelantan, Perlis, Trengganu, and the islands of Langkawi from Siam to Great Britain. The British, in turn, renounced extraterritorial rights in Siam and acknowledged Siamese sovereignty over the provinces of Patani, Narathiwat, Yala, and Satun.

9. A *pondok* is a traditional private Islamic school, offering both basic and advanced Islamic studies. Instruction is in Malay and, for advanced courses, in Arabic. It is normally owned and administered by a well-versed and respected traditional religious teacher.

2
History of Muslim Resistance Movements in the Twentieth Century

THE Muslim armed resistance movements in Mindanao and Sulu and in Patani have important places in the history of the Philippines and Thailand respectively. The Moros fought Spanish colonialism for over 300 years and resisted the military strength of the United States for almost half a century. More significantly, they have posed a serious challenge to the Philippine government through their guerrilla activities and organized liberation fronts, particularly since the late 1960s. Similarly, the Malay-Muslims of southern Thailand have persistently opposed the Thai control of the Patani area since it first became a vassal state of Thailand at the beginning of the seventeenth century. When Patani was finally incorporated into Thai territory three centuries later, a series of uprisings and armed movements against the Thai government followed. These provided the basis from which organized separatist fronts emerged at the end of the 1950s.

The Moro Resistance Movement

Resistance to the American Regime, 1899–1946

The transfer of sovereignty over Moroland from Spain to the United States did not render the Moros less vigorous in their resistance to colonialism. Thousands of them fought and died resisting the American policy of incorporating their homeland into the Philippine state.

Military Occupation (1899–1903)

This period represented the initial contact of American colonizers with the Moro people. The American efforts to gain Moro recognition of its sovereignty over Mindanao and Sulu and to keep the Moros out of the Philippine–American War (1899–1901) which was being waged in the northern islands of the Archipelago contributed to the signing of the Bates Agreement in August 1899 between Brigadier-General John Bates of the United States Army and Sultan Jamalul Kiram II of Jolo and four of his ruling aristocrats. The essence of the Agreement was that the Moros acknowledged American sovereignty and pledged to help suppress piracy and arrest individuals charged with crimes against non-Moros. In return, the United States agreed to protect Moros from foreign intrusions and to respect the authority of the Sultan and other chiefs.

At first, the Moros seemed to be content with these arrangements because the United States followed a policy of strict non-interference (Stuart C. Miller, 1982). After the Philippine–American War was over, however, American military leaders believed that it was necessary to implement a policy of direct control over Moroland.

To the Moros, the move posed a threat to the survival of their community. Resistance to the military occupation occurred in a number of places. For instance, Panglima Hassan and his men fought American soldiers in Sulu in 1901. Hassan's resistance was inspired by the belief that the continued American presence would adversely affect Moro authority, especially his own authority as a *panglima* (district chief) who commanded considerable influence among the *datu* of Jolo Island. Hassan was also convinced that resisting foreign domination was a *sabilillah* (struggle in the name of Allah) and used the concept to rally his supporters (Tan, 1977: 22).[1]

In Lanao, one of the more significant Moro armed struggles was the Battle of Bayan in May 1902. The conflict was caused mainly by the hostile attitudes of some of the Maranao sultans and chiefs towards American military activities. They accused the Americans of attempting to change their religion and to enslave their people. The Sultan of Bayan, assisted by the Sultan of Pandapatan, the Datu of Binadayan, and Moro fighters from various neighbouring *barangay*, such as Bacolod, Butig, and Maciu, engaged the 27th Infantry Company and the 25th Mountain Battery. More than 300 Muslims, including the Sultans of Bayan and Pandapatan, died in the battle. The American casualties numbered 10 killed and 41 wounded (Gowing, 1977: 86–7). There were other instances of Maranao resistance to the military efforts to bring Lanao under American rule. In June 1902 Datu Tungul and followers of Sultan Binibayan attacked troops near Camp Vicars, resulting in the death of the Sultan. A series of assaults by the Muslims followed. Between September and October 1902, Sultans Ganduli and Tanagan fought American soldiers; the Sultans and more than 100 Moros died (Tan, 1977: 21).

Resistance to the American military occupation was organized mainly by the Moro aristocrats who saw the American presence as a threat not only to their position and prestige but also to the Muslim community. The concept of *jihad*, therefore, justified their actions.

Moro Province (1903–1913)

A change of American policy from non-interference to direct control resulted in the establishment of Moro Province. Moro Province was organized and designed to make preparation for the integration of the Moro people into a future Philippine state, to which the Philippine Bill of 1902 was committed. The Province, divided into five districts—Zamboanga, Lanao, Cotabato, Davao, and Sulu—came under a policy of direct rule. President William McKinley in 1899 stated 'that the Philippines are ours, not to exploit but to develop, to civilize, to educate, to train in the science of self-government' (quoted in Harrison, 1922: 36).

As part of the programme to 'develop', 'civilize', and 'educate', the American system of government and concepts of justice were introduced. Well-ordered provincial and district governments were organized. Schools and hospitals on the Western model were built. Agriculture and commerce were expanded. Certain practices of the Moros, such as slavery, were made illegal. Furthermore, Filipino Christians from the northern provinces were encouraged to migrate to Moroland. It was also essential from the American point of view that the Moros be brought into the taxation system. The *cedula* (head tax) and other forms of levies, such as export–import duties and vessel registration fees, were imposed.

This policy of direct control unilaterally abrogated the Bates Agreement and resulted in disruption of the socio-political structure and customs by which the Moros had lived for centuries. The creation of provincial and district units of government whose officials enforced laws and regulations weakened the power and position of the Moro leaders. The American public school system undermined the Moro *pendita* schools. The parcelling out of lands to Christian settlers not only took from the Moros their major economic resource but also threatened their traditional practice of ancestral landholding. From the Moros' point of view, the creation of Moro Province meant the imposition of laws and customs of foreigners and of *kuffar* (infidels). It constituted a serious threat to their distinct Muslim *ummah*.

Moros under the Moro Province were administered by Generals Leonard Wood (1903–6), Tasker H. Bliss (1906–9), and John J. Pershing (1909–13). They were viewed by the American authorities as 'savages' whose laws and customs were not worth preserving. Thousands of Moros died fighting the American troops.

When Wood passed anti-slavery legislation and replaced the *agama* (religious) courts with the Western system of jurisprudence, Panglima Hassan again gathered his warriors and with the support of Sultan Jamalul Kiram II defied the government of the Moro Province in October 1903. The sporadic engagements of Panglima Hassan's men with the American soldiers continued until Panglima Hassan died in March 1904. Wood described Panglima Hassan's defiance as an expression of Muslim discontent.

Datu Usap, who strongly opposed the *cedula* and was also sympathetic to the cause of Panglima Hassan, resisted the authorities. Datu Usap's resistance was influenced by a religious preacher, Hajji Masdali, who persuaded him to rebuild his old *cota* (fort) and sold him *anting-anting* (charms) to render the fort and his people invulnerable (Gowing, 1977: 158; Tan, 1977: 29). In January 1905 Datu Usap rebelled. Major Hugh Scott, who had played a major role in defeating Panglima Hassan, assaulted Datu Usap's fort and ended the rebellion. The Datu died in the fighting and Hajji Masdali was captured and deported to Singapore.

Datu Pala, a supporter of Panglima Hassan who had retreated to British North Borneo, returned to Jolo Island and called for a *jihad* against the Americans. The call for a holy war was also triggered by Wood's anti-slavery policy, which deprived the Moro leaders of their privileges and sources of wealth. In May 1905 the American soldiers fought a series of battles with Datu Pala's forces. Datu Pala was killed but his men continued to resist until they were defeated and dispersed at the end of the year.

Datu Pala's relatives, supporters, and the *datu* who objected to American policies grouped together on Bud Dajo. The Battle of Bud Dajo between the American forces of about 790 men and more than 1,000 Moros occurred in March 1906 and lasted two days. More than 600 Muslims died (Gowing, 1977: 161). It was one of the bloodiest battles during Wood's term as Governor of the Moro Province.

In Lanao, there were frequent armed attacks on the American troops and on Christian settlements throughout the years 1903 and 1904. A series of campaigns by the then Captain John Pershing did not stop the disturbances. An area of especially strong resistance was Taraca. In April 1904 the American troops finally surrounded Taraca fort; many Muslims died and several powerful *datu* and *pendita* surrendered (Gowing, 1977: 154–5; Tan, 1977: 21). The lesson of the Taraca battle brought a temporary peace to Lanao.

In Cotabato, Datu Ali of Kudarangan, who not only refused to obey the anti-slavery law but also believed that the 'infidels' would force Muslims to become Christians, mobilized the people of the Cotabato Valley to revolt against the Americans. In March 1904 Wood's troops succeeded in destroying Datu Ali's main fort. However, Datu Ali changed his strategy from conventional to guerrilla war. The guerrillas constantly raided the Cotabato Valley, keeping the area in a disturbed state. Finally, a battle between Datu Ali's men and the 22nd Infantry Company under Captain Frank McCoy took place in October 1905 near Simpetan. Datu Ali and no less than 100 of his men were killed.

When General Tasker H. Bliss succeeded Wood in April 1906, the Battle of Bud Dajo had just ended and the scattered pockets of Moro resistance seemed to be exhausted. Bliss saw the opportunity to win the Moros by peaceful means and to move ahead in civil affairs. He wrote: 'The authorities forget that the most critical time is after slaughter has stopped. Then is when we need here men of influence and power to get the people started in the right way, to get them to cultivate their fields, and make them understand that peace is better for them than war' (quoted in Gowing, 1977: 170).

While governing Moroland with a firm hand, Bliss devoted his efforts to education of the Moros and other non-Christian tribes. He believed that the major means by which they could be brought into the mainstream of Philippine social and political life was education (Gowing, 1981: 17–21).

It would be misleading to suggest that Bliss did not value military power and see it as important in maintaining peace and order in Moroland. In fact, he increased the number of Moros in the Philippine Constabulary, feeling that Moro Constabulary officers and men were helpful in field operations against outlaws (Forbes, 1945: 109–11). During his three-year term as Governor of the Moro Province, there were no major acts of Moro resistance or rebellion. His period was known as the 'era of peace'.

The absence of major armed struggles, however, was not proof of Moro endorsement of American rule. Rather, it was because the Moros were exhausted after a period of vigorous resistance against Wood's administration. More importantly, Bliss's 'velvet glove' approach was less repressive than Wood's 'mailed fist' policy and required less vigorous resistance. As Faruki (1983: 9) suggests, 'Violent repression only moves ethnic movements to more and more extreme positions and greater radicalization.'

The relative peace in the Moro Province continued after General John J. Pershing became Governor in 1909. Pershing spent much of his time attempting to improve social and economic conditions in the province. His emphasis was on the development of the region's agriculture, especially to increase the variety and output of commercial crops. A special effort was made to develop the potential of Cotabato and Lanao, whose agricultural resources were largely underexploited. Pershing was also interested in developing mining and manufacturing.

In social affairs, Pershing's educational programme aimed at training the Moro children to embrace American social values and to be productive participants in the economic life of the region. More significantly, the General believed that peace and order were vital for social improvement in Moroland and that real peace would not occur until the Moros were disarmed.

The disarming of the Moros had been discussed during the terms of Wood and Bliss, but it failed to materialize. The arguments were: since the Moros identified the possession of weapons with their manhood, depriving them of their arms would mean violent resistance; most weapons in Moro hands were obsolete and used mainly for hunting; and it would be unfair to deprive the law-abiding Moros of their protection from lawless Moros. Pershing, however, felt that the disarming of the Moros was necessary, and it was time, he reckoned, 'to teach the Moros the meaning of government' (Gowing, 1977: 235). On 8 September 1911 Pershing issued Executive Order No. 24, effecting the disarmament of the Moros in the Moro Province.

Although Pershing promised payment for all arms surrendered, many Moros opposed the disarmament order. Clashes erupted between government troops and Moro armed bands who refused to turn in their weapons. In December 1911 more than 1,000 Moros took their positions on Bud Dajo in defiance of the government's disarmament policy. Because the incumbent Republicans feared that the Democrats would make political capital out of any major battle with the Moros, acting

Governor-General Newton Gilbert asked Pershing to call off his plan to assault the hostile Moros on Bud Dajo. Complying with Gilbert's wish, Pershing negotiated with the Moros and was able to persuade most of them to abandon the mountain and to return home. Some hardcore Moros fled into the jungle to continue their defiance, but they were captured later.

In Lanao, the Army and Constabulary units immediately dispersed the Maranao groups opposing the disarmament. Several thousand firearms were collected, including high-powered rifles. In Cotabato, Datu Alamada of Pedatan, a sympathizer of Datu Ali, and several hundred of his followers resisted the disarmament. His group was later driven into the interior, while some *datu* helped the government in the collection of arms.

In early 1913 the Moros of Lati Ward in Sulu, who were always hostile towards the disarmament policy, engaged the government forces in fierce *cota* battles. As more and more government troops came to fight them, nearly all of the population of Lati Ward, most of them unarmed villagers, retreated to Bud Bagsak. Negotiations between the government and the Moros on the mountain continued for months. They finally agreed that all Moros should return to their homes and turn in their arms. In return, the government would withdraw its troops from that area. Several Moro leaders, including Naqib Amil, Datu Jami, and Datu Sahipa, and their followers, however, refused to comply with the agreement. On 11 June 1913 General Pershing ordered his troops to fight them. The famous Battle of Bud Bagsak lasted five days. It is variously estimated that from 500 to 2,000 Moros were killed (Gowing, 1977: 238–42). With the exception of a few disturbances such as Datu Sabtal's resistance to tax collection in August and October 1913, peace returned to Moroland.

The Department of Mindanao and Sulu (1914–1920)

On 15 December 1913 the Moro Province was reorganized into the Department of Mindanao and Sulu to reflect the wish of the Democratic Party to speed up the movement towards Philippine independence in accordance with the principle of 'the Philippines for the Filipinos'. Frank W. Carpenter, a civilian, was appointed Governor of the Department. His main tasks were to unify Mindanao and Sulu with Luzon and the Visayas, to transfer the control of Moro affairs from Americans to Filipinos, and to continue the socio-political development of the Moro people with the aim of accelerating the course of their integration.

Because the ability of the Moros to resist had been weakened during the Moro Province period, Governor Carpenter was able to withdraw most of the United States Army units, except some garrisons of Philippine Scouts, from Moroland and leave the maintenance of peace and order to civilian responsibility. The unification of the region with the rest of the Philippines was accelerated by extending to the Department the jurisdiction of the Insular government. A policy of 'Filipinization' was pursued by giving Filipino Christians greater power in the administration of Moroland. Many Moros resented these changes and adhered tenaciously to their traditional ways while others yielded to the government programme of integration. The abdication of Sultan Jamalul Kiram II of Sulu from his temporal power in March 1915 was a classical example of Moro submission.

After President Woodrow Wilson signed Public Act No. 240 (the Jones Law)[2] in August 1916, the Philippine Legislature assumed legislative control over Mindanao and Sulu. Anxious to take over the American burdens in the affairs of the Moros, Filipino officials under Carpenter's supervision carried out 'the policy of attraction' which stressed the improvement of Moro welfare in order to further their integration (Ralph B. Thomas, 1971: 45–82).

Although the period of government under the Department of Mindanao and Sulu was generally peaceful, there were instances of Moro disturbances. In 1917, the *datu* of the Bayan area defied the government. They refused to turn in their arms, to permit their children to attend public schools, or to allow officials to survey their land. *Cota* fights between them and Constabulary and Scout units erupted. The hostile *datu*, *pendita*, and most of their men died in the battles. Their defeat strengthened the government control of Lanao. In October of the same year, Datu Ambang of Kidapawan in Cotabato gathered his people and invited several *datu* to carry out 'God's command' to fight the infidel Americans and Filipinos. When the Governor of the province was informed of the matter and ordered the government forces to meet any serious outbreak, the movement lost its momentum (Tan, 1977: 31–2). There were several other minor incidents and resistance activities against the government, but these were quickly suppressed.

The six-year administration of the Department of Mindanao and Sulu was a period of acceleration of Moro integration into a united, self-governing Philippines. With the abolition of the Department, the effective years of American administration in Moroland came to an end. The administrative and legislative control of the Moros, through the Bureau of Non-Christian Tribes, was firmly in the hands of Christian Filipinos.

The Bureau of Non-Christian Tribes (1920–1937)

Act No. 2878 of the Philippine Legislature formally abolished the Department of Mindanao and Sulu in February 1920. The administration of Moroland was assumed by the Department of the Interior, exercising its powers through the Bureau of Non-Christian Tribes. Though American officials continued to play important roles in administering the affairs of the Moros, the region was now under the control of the Insular government which, with respect to domestic matters, was largely in the hands of Christian Filipinos. The Philippine Legislature, for instance, had power to veto the appointment of provincial governors and other key officials made by the American Governor-General and to appropriate funds for programmes in Moroland.

Aside from certain leaders who were willing to co-operate with the Manila authorities for their own personal interests, the Moros as a whole preferred to live under the American government than to be controlled by Filipinos whom they considered their old enemies. For them, it was one thing to assent to the government of Americans who defeated them in many battles, but it was another to yield to the rule of Christian Filipinos who, as troops under the Spaniards, had never effectively defeated them despite attempts over 300 years (Gowing, 1979: 168). Moreover, the Moros believed that Christian Filipinos, influenced by centuries of Spanish domination, had hidden motives to stamp out their religion and traditions. The Filipinos also needed their land for economic and territorial expansion (Ralph B. Thomas, 1971: 133). In 1921 the Moro leaders in Sulu presented a petition to the President of the United States requesting that Sulu be governed separately from the rest of the Philippines. In 1924 another petition by a delegation of Moro leaders at Zamboanga was addressed to the United States Congress. It expressed the Moro desire that Mindanao and Sulu be made a territory of the United States (Gowing, 1979: 168–9). Their desire to be separated from the Philippine nation suggested that the centuries-old Moro–Filipino animosity, despite outward appearances and reassuring words, remained strong. The desire of the Moros went unheeded, however. Moro and Filipino communities were incorporated.

The Moros also expressed their discontent through armed resistance, though they no longer had the strength to offer the same threat they had to the Americans up to 1913. In Lanao, numerous clashes were reported between government troops and Moros. At Sisiman in 1921, Datu Amai Binaning and several of his followers died in a battle with Constabulary soldiers to oppose compulsory government education. In 1923 the Muslims of Tugaya organized armed uprisings in defiance of taxes, education, and exactions for road construction. Fifty-four of the Maranao Muslims lost their lives (Gowing, 1979: 170; Tan, 1977: 42). In the same year, the Moros in Ganassi took revenge against the brutality of government soldiers; several Muslims died in the engagements, including their leader, Saruang. Datu Pandak, a sympathizer of Saruang, defied the government in the following year. Datu Pandak's movement provoked both religious and political antagonisms and raised the fighting spirit of many Maranao people. But the movement disintegrated soon after Datu Pandak was killed (Tan, 1977: 32, 42).

In Cotabato in 1923, Datu Santiago and his men, some of whom were Constabulary deserters, attacked government troops around Parang. They objected to the *cedula*, the maltreatment of the Muslims by Philippine Constabulary, and the abuse of some school officials who had been forcing Datu Santiago to repair school facilities without compensation. The unrest lasted one year and cost more than 100 Moro lives. The Datu finally surrendered (Tan, 1977: 41–2).

The Acbara resistance erupted in Pata Island, Palawan, in May 1923. It was attributed to the *cedula*, compulsory education, and land surveys. Acbara, a trader, was also influenced by a religious preacher, Hatib

Sihaban, who wanted to drive out the Americans and Christian Filipinos in order to establish a government 'responsible only to God'. Acbara told his followers that it was God's wish that the Muslims resist the government. In June, his group fought government forces; he and 32 of his followers were killed. Another disturbance with a similar religious element was reported in 1924 in Bucas Island off the eastern coast of Mindanao. A man calling himself 'Imam Mahdi' (Messiah) convinced some Muslim villagers that their mission was to fight against foreigners and unbelievers. 'Imam Mahdi' and his people killed ten soldiers before they were suppressed (Tan, 1977: 32–3). In Balete, Agusan, seven soldiers were killed in an ambush by a group of Muslims related to the Mahdi's movement, but government reprisals put an end to the dissidents.

The United States Congress put on record in 1926 the petition sent to it in 1924 by Moro leaders who expressed their intention to declare themselves an independent Moro Nation should the United States grant independence to the Philippines without provision for retention of Moroland under American rule (see above). The Bacon Bill of 1926 also proposed that Mindanao and Sulu be retained under the American flag. In this political atmosphere, Datu Tahil, who was disappointed at not being appointed governor of Sulu, defied the government. He demanded the reduction of *cedula* and land tax, and the privilege of carrying weapons. In January 1927 his fort at Patikul was attacked by government troops, who killed 30–40 of his followers. Datu Tahil surrendered (Tan, 1977: 39–40).

Despite the attempts of Moros to resist integration, the official policy of the United States remained always to incorporate Moroland into the Philippines. To that end, the United States Congress passed the Tydings-McDuffie Independence Act in May 1934 to establish a Commonwealth in which Filipinos would function with full power over internal affairs. The Moros were decidedly against the formation of the Philippine Commonwealth, which they considered as the last step towards their dependence. In the same year, a series of armed incidents occurred in Lanao. Amai Milon and Dimakaling were among the leaders of the disturbances. They opposed the Constabulary practice of false arrest and the imposition of a fine on parents whose children were not in school. These armed bands were again defeated (Tan, 1977: 42). At the same time, 200 Moro leaders signed a letter drafted by a religious leader, Hajji Abdul Kamid Bogabong, and sent it to the Governor-General. They insisted, 'A law be passed that our religion, Islam, taught by Prophet Mohammad be not changed, or curtailed in any way and that we must not be forbidden to observe our religion' (quoted in Ralph B. Thomas, 1971: 253). In 1935 Bogabong and 119 Maranao leaders addressed another letter to President Franklin Roosevelt expressing their concern about the future of the Muslim community under the government of Christian Filipinos, particularly the maintenance of their religion and customs. They declared that 'All practices ... of Islam should be respected because these things are what a Muslim desires for.... Once our religion is no more, our lives are no more' (quoted in Glang, 1969: 17). Nevertheless, the Commonwealth government was inaugurated in November 1935. The regime was to be granted complete independence in 1946.

The Philippine Commonwealth (1935–1946)

After decades of resisting the American efforts to include their homeland in the Philippines, some Moro leaders realized that their resistance was pointless. Gradually, they accepted the new situation which they were powerless to change and sought to make the best of it. Arolas Tulawi of Sulu, Datu Menandang Piang and Datu Blah Sinsuat of Cotabato, and Sultan Alaoya Alonto of Lanao were elected delegates to the Constitutional Convention of 1934 by their Muslim brothers. With many votes coming from Christian constituents, the Moro provinces approved the new constitution. It was the first time the Moro people had participated in a Philippine nation-wide election. However, only two Moros, Datu Ombra Amilbangsa in Sulu and Datu Sinsuat Balabaran in Cotabato, were elected as Members of the National Assembly in 1935. They became the only voices speaking for the interests of Muslims in the Commonwealth. It is not surprising that the advancement of the Moros came last in national priorities for Moroland.

But President Manuel Quezon spoke of giving the Moros the best government and of his administration's devoted interest in their welfare and advancement (Ralph B. Thomas, 1971: 263–4). The Moros found it very difficult to identify specific government actions to this effect. On the contrary, their sense of alienation was deepened by the effort of the Commonwealth to put an end to the special treatment they had received under American rule. For example, the National Assembly repealed the Administrative Code for Mindanao and Sulu which had given the Moros certain exemptions from national laws, and the Moro Board, which had been established to settle some Moro disputes according to Islamic and traditional laws, was abolished. In 1937 the Bureau of Non-Christian Tribes, which had attempted to meet the special requirements of the Moros administratively, was replaced by the Office of Commisioner for Mindanao and Sulu (Gowing, 1979: 176–7). The Office's primary concern was the development of Mindanao for the immediate benefit of the Commonwealth as a whole (Ralph B. Thomas, 1971: 271). The Commonwealth government, furthermore, ended official recognition of the civil title held by Moros as expressive of their traditional social system. It also refused to acknowledge the successor to Sultan Jamalul Kiram II, who died in 1936. More significantly, the Commonwealth regime markedly reduced social and economic programmes specially designed for the Moros. The development efforts in Moroland were conducted largely by and for the benefit of the Christian settlers and foreign entrepreneurs (Gowing, 1979: 176–8).

Again, some Moros expressed their discontent towards the government through armed struggle. The more serious armed conflicts between Muslims and the Philippine Commonwealth were in the form of *cota* fights and occurred largely in Lanao during 1936 to 1941. Many of the scattered *cota* around Lake Lanao were destroyed. During the Japanese Occupation (1941–4), however, most Moros supported the anti-Japanese war effort (Ralph B. Thomas, 1977). This was due to the harsh treatment of their subjects by the Japanese, which offset the bad experience the Moros had had with the Americans and Christian Filipinos. It also suggests that the Moros were not anti-American or anti-Filipino *per se*; they resisted all forms of external power that threatened their community.

In sum, the armed resistance movements that occurred during the American regime, ranging from full-scale battles to minor incidents, were motivated by the presence of the Americans and Filipino Christians who were considered a threat to the position of Islam and the interests of the Muslims. The Muslims had always resisted any effort by non-Muslims to challenge their traditional position. Two other outstanding features of the Moro resistance should be mentioned. First, the Moro struggles were characterized by the absence of consistent and effective co-ordination and co-operation among different Muslim ethno-cultural groups and with other non-Christian tribes. This was explained by the fragmented nature of the Moro society and the mutual suspicion and distrust among the various ethnic groups. The rivalries and dissension among Muslim leaders and the kinship system which polarized loyalties and interests along blood lines prevented the Moro communities from forging a unified resistance to the powerful outside forces (Tan, 1977: 60–1; 1982: 61–9). Secondly, the less vigorous armed resistance during the periods of civil administration (1914–34) and the Commonwealth government (1935–46) compared to the era of the military regime (1899–1913) was due mainly to the fact that the Moros had been largely disarmed and pacified and were exhausted. In addition, the nature of the civilian administration was less repressive and more tolerable than that of the military regime. Hence, it received a less forceful response from the Moros.

Resistance to the Philippine Government, 1946–1968

Moroland was structurally integrated into the Republic of the Philippines which was proclaimed on 4 July 1946. The 'Moro Problem'[3] inherited by the new Republic was the transfer of allegiance from traditional community ties to a larger state as a new political entity. The Moros, who constituted a nationality distinct from, and older than, that of Christian Filipinos, possessed a strong sense of group consciousness. They continued to assert their identity as Muslims. Many refused to regard themselves as Filipinos. As Glang (1969: 21) wrote:

The term Filipino can only refer to a segment of our people who bowed in submission to the might of Spain. Certainly, the Muslims do not fall under the category of Filipino. Being a historic people, the Muslims therefore cannot but reject the generalization that the word Filipino applies to them as well. Because when the word Filipino is applied to a segment of our people, the implication is that the word Filipino was derived or at least named in honour of King Felipe II, from [sic] whom the Philippines was named after. The name or term Filipino can only be applied, therefore, for those people who were once subjects of Spain. This is a historical fact which cannot be disputed. In so far as the Muslims are concerned, the appellation Filipino does not have any meaning to them. That is the reason why they resent the idea of being just called a Filipino.

This problem of conflicting loyalties was further compounded by a deepening sense of religious awareness due to the resurgence of Islam and by a deepening sense of deprivation because of the government's policy of integration.

The Deepening Sense of Islamic Consciousness

One of the significant factors that weld Moros of different ethno-cultural groups together is their religion—Islam. It has a powerful social and spiritual hold on them and becomes the symbol of their selfhood. This is not to say that Islam has been able to forge an unfragmented Moro community, but it has become one of the major ruling forces in the historical development of the Moros, a force that motivated them to resist the Spanish *conquistadores* and the American soldiers.

The general resurgence of Islam in the Muslim world after the Second World War (Gowing, 1979; Mastura, 1980; Ayoob, 1981) had a significant effect upon the Moros, notably in enhancing an atmosphere of religious and cultural awareness. Since the early 1950s, Muslim preachers from different parts of the Muslim world have come to Mindanao and Sulu, on an official as well as an individual basis, to contribute to a strengthening of commitment to Islam in Moroland. This is because the majority of Muslims in South-East Asia, including the Moros, are seen as still adhering to certain pre-Islamic practices that are considered as not fully Islamic (see Gowing, 1969). At the same time, Islamic institutions and universities, such as Al-Azhar University in Cairo and the Islamic University of Medina, provided scholarships to young Moros. Table 2.1 shows the number of scholarships awarded to Moro students by various Muslim countries through the Philippine government during the years 1977 to 1981. In addition to these students, a number of scholarships were given to Moro students through non-government channels.

Moro leaders were invited to participate in the conferences and seminars of various Muslim bodies, while an increasing number of Muslim villagers performed their *haj* (pilgrimage) in Mecca. During the years 1978 and 1980, for example, a total of 4,498 Moros performed their *haj* (Pangandaman, 1983: 248).[4] As a result, Islamic consciousness among the Moros was manifested in a proliferation of Islamic institutions such as mosques, *madaris*, and Muslim associations. These Islamic institutions strengthen the sense of Moro nationalism and solidarity (Gowing, 1979: 187). The term 'Bangsa Moro' emerged as an identifying name for the native Muslims of Mindanao and Sulu. Some of the Moro Islamic

TABLE 2.1
Number of Moro Students Awarded Scholarships by Muslim Countries, 1977–1981

Country	Number of Students				
	1977	1978	1979	1980	1981
Egypt	10	17	7	1	–
Indonesia	–	4	1	–	–
Kuwait	3	2	4	2	3
Libya	1	–	4	–	–
Saudi Arabia	55	54	82	31	57
United Arab Emirates	–	–	4	2	3
Total	69	77	102	36	63

Source: Pangandaman (1983).

institutions served to secure international contacts and to obtain assistance from the Muslim world (Pangandaman, 1983: 130–2, 136).

In 1983 there was a total of 987 *madaris* and 3,095 religious teachers (*ustaz*) with 132,811 pupils in the thirteen provinces of the autonomous regions (Ministry of Muslim Affairs, 1983: 14).[5] Most of these *madaris* offer Islamic courses that involve instruction in the Quran, the performance of rituals, and the basic principles of Islam. There are a few Islamic institutions of higher learning that provide advanced courses in Islamic studies. The Jamiatul Philippine Al-Islami in Marawi City and the Philippine Muslim College in Jolo are examples of such institutions. Certain *madaris* and institutions maintain close contacts with educational centres abroad. These contacts enable them to obtain occasional financial donations and scholarships from foreign nations.

Some *ustaz* functioned not only as religious teachers in religious schools but as Muslim preachers (*du-ah*); most were recruited by the Islamic call centres of different Muslim countries, such as Saudi Arabia, Libya, and Kuwait, to preach the word of Islam in the Philippines. According to some preachers interviewed in Marawi and Cotabato Cities in November 1984, there were more than 100 Moro preachers, most of whom were graduates of universities in the Middle East, employed by various Muslim countries. Table 2.2 shows yearly salaries and allowances of Moro preachers working under various Islamic call centres.

The revitalization of Islam has helped not only to heighten the religious solidarity of the Moros but also to sharpen the sense of 'difference' between them and Christian Filipinos. Moros prefer to be called 'Muslim' to emphasize their belonging to a different religion. They refer to the Manila government as 'the Christian government' to which no true Muslim owes allegiance (Gowing, 1962: 62). They view their sultans and *datu* as representing an institution of Islam and interpret the government's refusal to recognize its authority as rejection of their religion. In a Preliminary Report of the Special Committee to Investigate the Moro Problem, the Philippine House of Representatives

TABLE 2.2
Salary of a Moro Preacher by Islamic Call Centre

Islamic Call Centre and Country	Estimated Annual Salary and Allowance (US$)
Al-Auqaf, Kuwait	6,000
Al-Awqaf, Qatar	3,000
Al-Azhar, Egypt	2,400
Darul-Ifta, Saudi Arabia	3,750
Darul-Maarif, Saudi Arabia	3,000
World Islamic Call Society, Libya	3,150
Rabitat Al-Alam Al-Islami, Saudi Arabia	3,750

Source: Interviews of several Muslim preachers in Marawi and Cotabato Cities, November 1984.

(1956: 68) suggested: 'Any move to deny the authority of his sultan, or to curtail his freedom as an individual in a Muslim state would make him believe, in his own ignorant ways, that it is an affront on his religion. And he does not hesitate to do away with such curtailment even if it must cost his life.' The Moros, furthermore, became more articulate in defence of the faith as they became more knowledgeable about the doctrines and duties of Islam. Their concern was the preservation of their community and the elevation of their Muslim identity. They remained little interested in national goals, despite the efforts of the government to promote 'Filipinism' among them. Moro leaders such as Senator Domocao Alonto of Lanao del Sur developed close associations with Muslim organizations beyond the Philippines and Muslim politicians were speaking about forming an Islamic party (Gowing, 1979: 187). This growing awareness of the need to defend the Muslim community and concern about Christianization, prompted by religious missions and secular institutions which promoted Christian culture, were among the factors that triggered the Kamlon Rebellion (Tan, 1982: 68).[6] Another factor contributing to the Rebellion, as revealed by Kamlon to Secretary of National Defence Ramon Magsaysay, was fear of changes which tended to benefit the Christian immigrants rather than the natives (Tan, 1982: 116). These changes deepened the Moro sense of deprivation.

The Deepening Sense of Deprivation

As a young Republic, the Philippines was facing many challenges to nation-building. They included, for example, establishing economic viability, recovering from the devastation of the Second World War, and coping with the predicaments of minorities.

A multi-faceted policy was devised by the Philippine government in the early post-independence period to promote social and economic development and to render permanent, political integration of the Moros. The instrument for carrying out this policy was the Commission on National Integration (CNI), established by the Congress in 1957 under Act No. 1888. The accomplishments of the CNI, however, fell far short of its objectives for reasons which included mismanagement of funds and suspicion of the government's real motives. The CNI was never well received by the Moros, who feared that the true goal of integration and development was the destruction of their Muslim identity (May, 1984: 429; cf. Tamano, 1969). The government's limited concessions to the religious and cultural demands of Moro nationalism were seen as temporary expedients.

Thus, Moro expectations about the benefits of participating in Philippine society as Muslims (Suhrke and Noble, 1977: 181) were shattered by their perceptions of real and imagined social and economic deprivation, increasing political disadvantage, competition with Christian migrants for land in their homeland, interference through the heavy hand of the government in local affairs, and fear of Christianization (Gowing, 1979: 187). Such perceptions contributed to a deepening sense of alienation among the Moros and to the persistence of the resistance struggles and unrest in the region.

In Sulu, for instance, the disturbances which were common in the post-war period were stimulated by factors related to socio-economic ills. The restriction of trade with Borneo, the occupation by Christians of available agricultural land, and strong competition from Christians in fishing resulted in limited opportunities for employment (cf. Philippine Senate, 1963: 5). It forced some desperate Muslims to turn to illegal employment such as smuggling and banditry. As the government attempted to suppress them and other rebellious Muslims, complaints of military abuses against innocent civilians became common. Moro resentment against the government for its general neglect of the province and the sense of deprivation intensified. In 1961 Datu Ombra Amilbangsa introduced the Congress House Bill 5682 calling for 'Granting and Recognizing the Independence of the Province of Sulu' (Gowing, 1979: 188). The Bill produced no concrete result, but it symbolized the Moro discontent.

In Mindanao, migration and competition for land had been the major elements contributing to unrest. The Philippine government sought to relieve severe population pressure in Luzon and the Visayas by encouraging migration to Moroland (for an account of resettlement programmes, see Silva, 1979: 39–55; Gowing, 1979: 189–91). As Table 1.2 indicates, the non-Moro population in Mindanao increased from 2,010,223 in 1948 to 6,294,224 in 1970. The policy also offered solutions for the political and economic difficulties of the north. For example, the resettlement of the landless Huks in Mindanao under the Economic Development Corps (EDCOR) programme was partly to disengage the Hukbalahap Rebellion in Luzon (see Kerkvliet, 1977). The concessions made to corporations and individuals for plantation production of export crops, logging, and mining benefited the national economy but brought little benefit to the Moros.

On the contrary, migration, the occupation of lands by outsiders, and the government's land laws had serious negative social and economic consequences for the Moros. The newcomers were active commercially and politically. The economic disparity between Moros and Christian settlers became apparent. The communities of settlers became centres of trade linked by arteries of good roads, while the Moro communities remained isolated (Gowing, 1979: 189). The Philippine Senate Committee on National Minorities (Philippine Senate, 1971: 22) reported that up to 1971 there was not a single irrigation project in any municipality where Muslims constituted a majority. In the words of George (1980: 122): 'Two decades after the Philippines became independent, Muslims in Mindanao were a devitalized people, their economic condition stagnant, their social traditions in jeopardy, their laws and customs in danger of disintegrating.' Faced with this threat, some Moros resorted to violence, and 'little wars' between the Christian settlers and the Moros mushroomed. One powerful force was a philosophy that was exclusively economic; another was deeply social–religious (George, 1980: 120).

The conflict in Moroland deepened after Ferdinand Marcos became President in 1965. The Moros were bitter about the so-called 'Jabidah Massacre' (see Chapter 3) in March 1968 and viewed it as demonstrating the low regard the Marcos administration had for their lives. In May, Datu Udtog Matalam announced the formation of the Muslim Independence Movement (MIM). His avowed objective was the creation of the Islamic Republic of Mindanao, Sulu and Palawan (MIM, 1968: 1). The motives attributed to the organization of MIM included Matalam's anger at the Corregidor Incident (Jabidah Massacre) and his aversion to the deteriorating general socio-economic condition of the Moros (Noble, 1976: 408–9).

In the final analysis, the Moro resistance that occurred during the American regime may be seen as a natural response to foreign domination. 'Foreign domination', states Smith (1971: 66), 'is always perceived to be oppressive.' Apart from political and economic oppression, the struggles were motivated by the desire to preserve the Muslim community as a distinct entity having its own territory, political institutions, and ethno-cultural integrity. Ibn Taymiyyah, a Muslim philosopher of the thirteenth to fourteenth centuries, suggested that 'the Muslims must maintain their distinct identity as a religious community and take extreme care not to merge themselves into other religious groups by imitating or associating themselves with their ways, customs, festivals, beliefs, etc.' (Qamaruddin Khan, 1973: 36). For Islam, which is the guiding force of the community, is the *raison d'être* of Muslim existence as a people. As a religion or ideology, it links Muslim lives to some larger and more enduring purpose and gives meaning to their existence. Any demand that this ideology be surrendered is intolerable. 'It would represent a kind of psychological death harder to contemplate than biological death' (quoted in Gowing, 1977: 323). It is this concept of 'psychological death' and the *jihad* which promised Muslims great rewards in the hereafter that rendered the Moros less discouraged by their defeats and loss of lives in defending the *ummah*.

The Malay Resistance Movement

Resistance to Thai Integration Efforts, 1899–1947

From 1899 onwards, Thailand under King Chulalongkorn (Rama V) gradually integrated its tributary states and outer provinces into the *thesaphiban* system of provincial administration (see Chapter 1). The partial integration of some of the tributary states such as the Seven Malay Provinces (*khaek chet huamuang*) was not satisfactory to Thailand. It wanted to exert more control over such states in order to secure the Kingdom's territorial integrity against increasing pressure from France and Great Britain. Greater control was also desirable to ensure the provinces' loyalty and to exploit more effectively their revenue and manpower resources (Bunnag, 1977: 137).

When the newly created Ministry of the Interior (*mahatthai*) issued the 'Regulations Concerning the Administration of the Area of the Seven Provinces', and when it became clear the position of local rulers (*chaomuang*) would be replaced by Thai appointees, the Malay *raja* turned to their Malay colleagues under the British for help. In 1901, Tengku Abdul Kadir Qamaruddin, Raja of Patani, appealed to Sir Frank Swettenham, the British Governor of the Straits Settlements in Singapore, for intervention in Patani's grievances. London, however, chose not to interfere in Thailand's policies in Patani. Any interference from the British would have tarnished Thailand's role as a buffer state between French Indo-China and British India. Moreover, the British had a plan to include the Malay states of Kedah, Trengganu, Kelantan, and Perlis, which were Thailand's dependencies at the time, into their domain (Pitsuwan, 1982: 31). The British naturally did not want to antagonize Thailand over the affairs of Patani.

Failure to obtain support from Britain did not discourage the Malay leaders of Patani from continuing their opposition to the concerted effort of the Minister of the Interior to centralize the existing administration. The resistance was co-ordinated by Tengku Abdul Kadir Qamaruddin, who attempted to organize a movement that could draw more oppressive measures from the Thais in order to spark violent uprisings against the policy of reorganization of the Seven Provinces (Pitsuwan, 1982: 51).

The first collective resistance movement began when the Malay *raja* refused to concede their power as *chaomuang* of the Seven Provinces. They ordered their subordinates to boycott all meetings with Thai authorities and directed the newly assigned Thai-Buddhist bureaucrats in the provinces not to perform official functions. Joining the nobility in defying Bangkok were Muslim religious leaders who believed that submission to a non-Islamic regime without resistance was not permissible in Islam. In Saiburi, for example, religious leaders supported Tengku

Abdul Kadir Nilebai (Phraya Suriyasunthornbowornphakdi), Raja of Saiburi, in protesting against the religious court being set up by the Thai authorities. In Raman, religious judges and village headmen resigned in protest against the encroachment of Bangkok. The Thai government had to resort to force to restore order (Pitsuwan, 1982: 54–5).

Muslim resistance to Thai integration efforts failed, however. Tuan Tengah Shamsuddin, Raja of Legeh (Ra-ngae), was deposed from his position in 1901 and was exiled to Songkhla. The deposition of other *raja* followed in 1902, but they were given fixed pensions as compensation. All functions of the Seven Provinces were gradually transferred to Thai bureaucrats (Haemindra, 1976: 203). The traditional state ritual of sending *Bunga Mas* to the Thai king every two and a half years to affirm their loyalty was no longer required because the Malay Provinces were then considered an integral part of the Thai state. Later, they were regrouped into four southern provinces.

In 1903, Tengku Abdul Kadir Qamaruddin, the former Raja of Patani, was arrested and sent to prison in Phitsanulok Province. He was released after spending 33 months in prison and pledging to refrain from politics. Tuan Tengah was permitted to return to Ra-ngae after having promised to comply with the government 'Regulations' (Haemindra, 1976: 202–3). Despite their pledges, Tengku Abdul Kadir Qamaruddin and his associates managed to organize a series of resistance movements and uprisings in the region.

Aside from the political transformation of the Seven Provinces from tributary municipalities to ordinary Thai provinces, the most important measure towards integration taken by Bangkok in this period was the replacement of the *sharia* and *adat* laws with Thai laws, except in matters concerning marriage and inheritance. Whereas administrative and financial measures caused strong resentment among the former ruling class whose power and interests were directly affected, the abolition of the *sharia* and *adat* stirred the deepest feelings of the Muslims in general, for every Malay-Muslim considers that his religious commitments, which are based on the *sharia* and customs, are 'the axis of his whole existence, his faith is what he lives for and would quite willingly die for' (Geertz, 1968: 111).

During the reign of King Vajiravudh (1910–25), the resistance movements intensified as the effect of Thai control over Malay society increased. Most of the unrest was essentially motivated by the desire to regain self-rule as a Muslim community. Indeed, from the very beginning of the incorporation of Patani into the Thai kingdom in 1902, Muslim resistance took both political and religious forms. The involvement of religious leaders in the struggles to regain the political power of the nobility was a case in point. Rebellions led by charismatic religious leaders—To' Tae in 1910 and Hajji Bula in the following year—were further instances of religious hostilities (Ministry of the Interior, 1967: 8).

In 1915, Tengku Abdul Kadir Qamaruddin retreated to Kelantan, which by then was under British rule (Mohamad, 1974: 39–40). From there he continued to provide leadership for resistance movements with the assistance of the Sultan of Kelantan, Sultan Muhammad IV. The most serious uprising inspired by Tengku Abdul Kadir Qamaruddin's efforts to resist Thai rule occurred in 1922 after the Thai government had taken another major step towards Muslim integration by promulgating the Compulsory Primary Education Act in 1921. The Act required all Malay-Muslim children to attend Thai primary schools. It was introduced to promote the use of the Thai language among the Muslims (Haemindra, 1977: 92).

The Malay-Muslims regarded this Act as part of a programme of 'Siamifying' the Patani people and stamping out their religion and culture. To them, it was crucial that their young children should not be exposed to education that would divert their attention from the teachings of Islam. As a reaction to the government's enforcement of compulsory Thai education and to the alleged funnelling out of the area of a large share of the revenues collected from the Malays, the former Malay nobility and some religious leaders ordered the Muslim villagers of Ban Namsai, in Mayo District, not to pay taxes and rent on land to the Thai government. In 1922 the villagers of Ban Namsai clashed with the military and police forces. The rebellion resulted in many casualties and the execution of several Muslim leaders (Fraser, 1966: 203; Pitsuwan, 1982: 57–8). In the following year, Thai authorities were accused of closing the Muslim vernacular and Quranic schools, resulting in another outbreak of protests. These resistance movements required substantial Thai forces for their suppression.

The outbreaks in 1922 and 1923 forced Bangkok to reassess its integration policy. King Vajiravudh issued new guide-lines for dealing with the Muslims in an attempt to lessen conflict in the provinces. These were aimed mainly at redressing some of the government regulations and practices that appeared inconsistent with the teachings of Islam. Taxation of the Muslim villagers was also minimized. Bangkok became aware of the emerging sense of Malay nationalism among the people in the northern states of Malaya and of their willingness to extend support to their brethren across the border. More importantly, it was perceived that there was a real threat of losing Patani to Britain if policies for the political and cultural integration of the Muslims were carried out indiscriminately.

The government's approach of cultivating political loyalties through sustained efforts of economic development and political participation, particularly after Thailand underwent a constitutional change in 1932, gradually created a sense of national belonging among the Malay-Muslims. Muslim willingness to participate in the Thai national political arena was demonstrated by the fact that three of the four Members of Parliament elected in the four Malay provinces in the general election of 1937 and 1938 were Muslims (Siamban, 1985). Another indication was the return to Thailand from Kelantan of Tengku Mahmud Mahyuddin, the youngest son of the former Raja of Patani.

Some Muslim leaders believed that they could gain concessions from the government and maintain their Muslim identity by participating in the existing system, while others, who were doing so, sought to buy time until their bargaining power increased. The Patani leaders realized that they could no longer depend on the support of their colleagues in the northern states of Malaya, due to the political control achieved by the British over them. In addition, Tengku Abdul Kadir Qamaruddin, who had been the central figure in rallying support for Patani since 1915, died in 1933. Indeed, a 15-year period (1923–38) of less vigorous pursuit of cultural suppression by the Thais was matched by correspondingly less forceful resistance by the Muslims.

Under the ultra-nationalist government of Phibun Songkhram from 1938 to 1944, the relatively peaceful situation in the former Muslim provinces ended. 'The posture of restrained participation adopted by the Malay-Muslims since the establishment of the parliamentary system had now turned to a policy of overt confrontation' (Pitsuwan, 1982: 90), as the government promulgated the Thai Custom Decree (*Thai ratthaniyom*) in 1939 in an attempt to change the cultural practices of the minority communities and to refashion the social habits of the entire population. Under this decree, the Malay-Muslims were forbidden to dress in the Malay fashion. More significantly, the use of the Malay language and certain practices of Islam were outlawed (Fraser, 1966: 203). In some cases, Muslims were compelled to worship Buddhist idols (Na Saiburi, 1944; GAMPAR, 1948: 3).

Phibun's policy of forced assimilation resulted in widespread resentment among the Muslims. Thousands left the region to find refuge in the Federation of Malaya and in Saudi Arabia. The refugees comprised largely the former ruling house and religious leaders (GAMPAR, 1948: 6). After Thailand joined the Japanese and declared war against the Allies in 1942, Japan placed the former Thai dependencies, Kelantan, Trengganu, Kedah, and Perlis, which were ceded to Britain in 1909, under Thai administration. The transfer brought immediate benefit not only to Thailand but also to Patani. The Muslims of Patani were once more able to renew their ties and share their problems with their brethren.

Tengku Mahmud Mahyuddin took the opportunity to carry on his father's struggle. He returned to Kelantan to organize the Muslims of Patani who had become refugees in Malaya. At the same time, he joined the British and Malay anti-Japanese efforts in order to gain support from them. In 1944 Tengku Abdul Jalal Tengku Abdul Muttalib, a Member of Parliament and son of the late Raja of Saiburi, submitted letters of protest to the Phibun and Khuang governments regarding the Thai policy of forced assimilation (Na Saiburi, 1944). Disappointed with the result of his efforts, he left for Kelantan and joined Tengku Mahyuddin's struggle.

At the close of the Pacific War in 1946, some Patani leaders hoped that Thailand would be treated as a defeated belligerent by the Western Allies and that the former Muslim provinces of Patani, Yala,

Narathiwat, and Satun might consequently become part of British Malaya. With that hope, Tengku Mahmud Mahyuddin, Tengku Abdul Jalal, and other Muslim leaders formed Gabungan Melayu Patani Raya or the Association of Malays of Greater Patani (GAMPAR) in Malaya. At about the same time, Hajji Sulong bin Abdul Kadir bin Muhammad Al-Fatani, a renowned religious teacher and President of the Provincial Council for Islamic Affairs (Majlis Agama Islam), also organized the Patani People's Movement (PPM) in Patani. This was the beginning of a long difficult struggle for autonomy and independence that has persisted to the present day.

Struggle for Autonomy after the Second World War, 1947-1959

The attainment of independence by some subject peoples of South-East Asia after the Second World War had aroused nationalistic aspirations among the Muslims in the region. The Patani Malay nationalists who resented Thai control over the four provinces organized GAMPAR in British Malaya with its principal offices in Kelantan, Kedah, Penang, and Singapore. Although GAMPAR was headed by former princes, with Tengku Mahmud Mahyuddin as its President and Tengku Abdul Jalal as its Deputy President, its aim was no longer to fight for the restoration of power to the former indigenous rulers. GAMPAR's main objective was to strive for irredentism of Patani with the Malay states on the peninsula. As the *Sunday Tribune* (14 March 1948) reported, 'Hundreds of Malay refugees from the southern provinces had held a meeting in Kota Bharu, Kelantan, and had unanimously decided to demand the separation of the four southern provinces from Siam and then linking them up with Malayan Federation.'

In Patani, Hajji Sulong, Hajji Wae Semae Muhammad (President of a Muslim association of Patani called 'Semangat' ('Zeal')), and 100 other Muslim leaders mobilized the Patani Muslims under the banner of the PPM (Bangnara, 1976: 106). The PPM was linked with GAMPAR and Semangat and was supported by different groups of Muslims. This was the first time that the leadership of a resistance movement of such magnitude was dominated by the religious élite. It marked an important change from the earlier struggle, which was led largely by former aristocrats.

On 3 April 1947 the PPM, under the leadership of Hajji Sulong, presented a set of demands to the government of Thavan Thamrongnavasawat (Whittingham-Jones, 1947). Its objective was to lay the foundation for an autonomous Patani state consisting of the four Muslim provinces. On 9 September of the same year, incidents flared up in Kampung Belukar Samok, Bacho District, Narathiwat. Police units burnt the village, causing forty families to lose their homes (Bangnara, 1976: 100-1). On 16 January 1948, Hajji Sulong and his associates, including Hajji Wae Semae Muhammad, were arrested and charged with treason. Their arrest produced a strong reaction among the Malay-Muslims and, in February, uprisings erupted in several places; in one of these, eight policemen were killed in a clash with Muslim villagers near Kampung Resab, Patani (*Straits Times*, 5 and 10 March 1948).

The situation in Patani continued to deteriorate when Phibun returned to power on 8 April 1948. On 26 April 1948, a revolt occurred in Kampung Dusun Nyor in Ra-ngae District, Narathiwat. The revolt was believed to have been instigated and organized by a religious teacher, Hajji Abdul Rahman (To' Paerak), and some nationalist leaders who had fled from the destroyed Kampung Belukar Samok. These instigators included Hajji Mat Karang, Che Senik Wan Mat Seng, and Zakaria Lalo. The fight against Thai police forces, which lasted two days, cost many lives. It was estimated that 400 Malays and 30 policemen were killed in the battle (Bangnara, 1976: 120). It resulted, moreover, in the flight of some 2,000–6,000 Muslims to Malaya (*Straits Times*, 29 and 30 April 1948). The Dusun Nyor Revolt, as it was known to the Malays, was the most serious uprising since the arrest of Hajji Sulong. Although the Thai government succeeded in suppressing the revolt, the political situation in the provinces remained critical. An estimated 250,000 Patani Muslims signed a petition requesting the United Nations to preside over the separation of the four Muslim provinces and its subsequent irredentism with the newly formed Federation of Malaya (*Far Eastern Economic Review*, 20–26 June 1980). Faced with this situation, the Phibun government declared a state of emergency in the Muslim area and sent three regiments of special police to Narathiwat, ostensibly for the purpose of 'combatting the Communists' (*Singapore Free Press*, 28 July 1948).

The Muslim leaders who fled to Malaya had rallied behind Tengku Mahmud Mahyuddin, the leader of GAMPAR. For his part, Tengku Mahmud Mahyuddin made a series of appeals to the United Nations, urging the Security Council to investigate Thai administration in Patani and to organize a plebiscite (Haemindra, 1976: 214). Strong protests were sent to various international organizations, such as the Asia Relations Organization and the Arab League. Calls for support were addressed to Muslim countries, such as the states of the Arab League, Indonesia, and Pakistan (Haemindra, 1976: 224).

Meanwhile, Muslim unrest in the region received increasing publicity in the Malayan press. The Muslim demands for autonomy or irredentism with the Federation of Malaya were supported by active Muslim sympathizers in various states in Malaya. On the other hand, Britain denounced the Patani Muslim demands and, in November 1948, raided GAMPAR's office in Singapore (Suhrke, 1975: 197). British Prime Minister Clement Attlee, who was eager to increase Thai rice exports, was not willing to risk offending the Thai government. He contented himself with making it understood that Britain expected a just solution to the Patani case (Thompson and Adloff, 1955: 161; Whittingham-Jones, 1948). In January 1949 an Anglo-Thai Agreement for joint control of the border was made. With this tacit agreement, Suhrke (1975: 197) concluded, 'The irredentist–separatist conflict was effectively neutralized on the intergovernmental level.'

The long-drawn-out trial of Hajji Sulong and his group ended on 24 February 1949. The court dismissed charges of treason, but imposed a 7-year prison sentence upon them for slandering the government in pamphlets distributed to the local population (Haemindra, 1976: 224). However, they were imprisoned for only 42 months and were released in 1952 (Thailand, Ministry of the Interior, 1967: 10). Two years later, Hajji Sulong, Hajji Wae Semae, and Che Sahak Yusof were requested by the Police Intelligence Chief in Songkhla, Pol. Lt.-Col. Bunlert Lertpricha, to report to him for consultation. On 13 August 1954 the three, together with Hajji Sulong's eldest son (as interpreter), went to Songkhla as requested; but they were killed in Songkhla (To' Mina, 1980–1).

With the condemnation of GAMPAR by Britain, the Anglo-Thai cooperation for joint control of the border, the assassination of Hajji Sulong and his group, and, above all, the failure of the Malay leaders (whom Britain accepted as representatives in the 1946–7 negotiations leading to the independent Federation of Malaya) to endorse the Patani case, the opportune moment for incorporation of Patani into Malaya seemed to have passed. This caused the Patani leaders and their Malay sympathizers to modify the irredentist aim. On the one hand, the irredentist problem would seem to have been contained. On the other hand, some Muslim nationalists and religious leaders felt that their goal had been modified and upgraded to a higher target with the hope of restoring the independence enjoyed by Patani in the olden times. Their new goal was also stimulated by a deepening sense of Islamic consciousness due to the resurgence of Islam.

The Deepening Sense of Islamic Consciousness

Like the Moros, the Malay resistance struggle was enhanced by an atmosphere of religious and cultural awareness. Such atmosphere was stimulated by the resurgence of Islam in the Muslim world and by Patani's historical role as 'a cradle of Islam' in South-East Asia. Patani was perhaps the most important region in the Malay peninsula in the nineteenth and early twentieth centuries for religious education and scholarship and had the largest and most famous *pondok* (Winzeler, 1974: 266). There are no records establishing when the first *pondok* in the region was founded, but among the earlier and more well-known *pondok* were Pondok Dala, Pondok Chuwuk, Pondok Berming, Pondok Bendang Guchil, Pondok Bendang Daya, Pondok Semela, Pondok Chouk, and Pondok Kuala Bekah (information from interviews with several *tòk guru* and *ustaz* in Patani and Narathiwat provinces, May 1985).

Most of the earlier *pondok* were established by religious scholars who had been trained at one of the three famous *masjid* (mosques): Al-Haram in Mecca, An-Nabawi in Medina, and Al-Azhar in Cairo. These mosques, apart from being places of worship, functioned also as institutions of learning where selected *ulama* (learned men) taught everything from the fundamentals of the Islamic religion to doctrine, Sufi mys-

TABLE 2.3
Number of Registered Mosques, Religious Schools, Religious Teachers, and Students in the Patani Region

Province	Mosques (1985)	Religious Schools (1984)	Religious Teachers (1984)	Students (1984)
Patani	450	82	429	16,132
Narathiwat	382	55	393	11,161
Yala	236	50	303	9,543
Satun	118	15	90	3,251
Total	1,186	202	1,215	40,087

Sources: Thailand, Education Region II (1984); Siamban (1985).

ticism, history, and Arabic (see Abdullah Al-Qari Hajji Salleh, 1974: 89). It is believed that *pondok* were founded to follow the education system of these acclaimed Islamic institutions. Apart from religious education, the objective of the *pondok* was to inculcate morality based on Islamic principles, such as the principle that every Muslim has a duty to cultivate rationality (*akal*) and to avoid irrationality (*nafsu*). Thus, *pondok* were viewed as an institution which provided both Islamic knowledge and moral guidance. Furthermore, *pondok* education was a basic requirement for all aspiring to become religious scholars, persons who are influential and respected within the Patani Muslim community.

With the resurgence of Islam after the Second World War, *pondok* and mosques in Patani have increased in number. Although they are subjected to certain government controls (see Chapter 5), they play a dominant role in strengthening the sense of Islamic consciousness and Malay identity. Table 2.3 shows the number of mosques and *pondok* in the four Muslim provinces.

In addition, many young villagers were sent abroad to continue their religious education. Some received scholarships from Islamic institutions in the Middle East. Meanwhile, more and more Muslim villagers made their pilgrimage to Mecca, enhancing their Muslim identification as they become *hajji*. In contrast to the situation in Mindanao, however, very few foreign Muslim preachers came to Patani, and fewer funds were provided by outside contributors. This was due to two factors: some Muslim countries were more satisfied with the level of Islamic education in Patani than they were with that of Mindanao; and, secondly, Muslim leaders in Patani were less well known internationally because most of them were not recognized by Thai authorities. The resurgence of Islam, nevertheless, had raised religious consciousness among Muslims in the region and sharpened the sense of 'difference' between Malay-Muslims and Thai-Buddhists. Like the Moros, the Malays became more articulate in defence of Islam and more concerned with the preservation of their community. This was illustrated by Hajji Sulong's famous demands presented to the Luang Thamrong government in 1947. The demands were:

1. The appointment of a single individual with full powers to govern the four provinces ... this individual to be local-born in one of the four provinces and to be elected by the people. 2. Eighty per cent of government servants in the four provinces to profess the Muslim religion. 3. Malay and Siamese to be the official languages. 4. Malay to be the medium of instruction in the primary schools. 5. Muslim law to be recognized and enforced in a separate Muslim court other than the civil court.... 6. All revenue and income derived from the four provinces to be utilized within them. 7. The formation of a Muslim Board having full powers to direct all Muslim affairs under the supreme authority of the head of State mentioned in (1) (Whittingham-Jones, 1947).

During 1958–60, demands for full independence of the region was justified by referring to the grandeur of Patani history and the teachings of Islam. Books to this end, such as *History of the Malay Kingdom of Patani* by Ibrahim Shukri and *Light of Security* by Hajji Sulong, were published. Leaflets were widely distributed to the villagers to raise their political consciousness. Indeed, the provinces continued to be beset by Muslim activists whose objective was no longer to become part of the Federation of Malaya but to gain complete independence. By the end of 1959, a national liberation movement was organized by Tengku Abdul Jalal, the former deputy head of GAMPAR (BNPP, 1981a: 5). The organization was called Barisan Nasional Pembebasan Patani (BNPP) or National Liberation Front of Patani.

The Moro and the Malay Resistance Movements Compared

An examination of the history of Moro and Malay resistance movements reveals some striking similarities. For one thing, the two had comparable political objectives: to resist the incorporation and integration of their communities into the Philippine and Thai political systems. The roots of resistance lay in the establishment of Muslim sultanates in Sulu in 1450 and in Patani in 1457, which led to the emergence of new, distinct Muslim communities. The Moro resistance to Spanish subjugation, the 'Moro Wars', began in 1565 and lasted till 1898, when the American conquest of Mindanao took place. The Patani resistance to Thai efforts to subjugate it also started at the close of the sixteenth century and continued until the end of the nineteenth century, when Patani was formally incorporated into the Kingdom of Thailand.

There were, however, differences in the circumstances of the two areas. While Spain succeeded in containing Islam in the Philippines, in Patani Islam continued to flourish. Though Thailand managed from time to time to reduce Patani to vassalage, Patani was considered to be one of the Islamic centres in South-East Asia in the nineteenth century. This is partly because Thailand's interest was mainly in the expansion of its political domain, whereas the goals of the Spaniards in the Philippines were to gain both territory and religious converts.

There were also differences between the two struggles in that the Moro resistance movements, especially in the twentieth century, often confronted both the government and separate local Christian forces. The

Malays, on the other hand, seldom engaged with forces other than those of the government. This may be explained by the social dichotomy and enmity between the Christians and the Muslims in Mindanao which resulted from Spanish and American colonial policies, particularly migration policies during the American and the Philippine rules. Thus, the Moros, in comparison to the Malays, seemed to face larger forces and more protracted battles throughout the history of their resistance.

The Moros and the Malays perceived their resistance efforts as both political and religious, and thus political and religious leaders were often involved in the movements. Moreover, both leadership groups believed that their involvement was to comply with the teachings of the Quran and Hadith (the Prophetic traditions).[7] In Thailand, religious leaders gradually dominated the Patani resistance movements as the former nobility lost their influence among the Malay villagers. The traditional political leaders in Mindanao, however, remained strong and continued to lead the resistance.

During the first quarter of the twentieth century, the Moros fought to retain their power and to prevent their incorporation into the Philippine state, while the Malays strove to regain the power they had lost by virtue of the Anglo-Siamese Treaty of 1909. It was not until after 1920 that the government of Moroland was under the Christian Filipino administration created by the American 'Filipinization' policy, and the actual inclusion of the Moro community into the Republic of the Philippines occurred in 1946. This time factor is not insignificant. The longer the community is incorporated, the greater the opportunity the government has to cope with its integration problems. Perhaps it is because of the time factor, among others, that Thailand appears to have been more successful in dealing with the Malay resistance activities than the Philippines was with the Moros.

The deepening sense of deprivation experienced by the Moros was not paralleled in Patani. The formation of the BNPP at the end of the 1950s was not so much a reaction to Thai government actions against Patani but, rather, a result of nationalistic aspirations aroused by the neighbouring Malay nations, Indonesia and Malaysia, which gained their independence in this period.

The ability and capacity of the Moro and the Malay communities to resist integration were affected substantially by external intervention and by the internal characteristics of the Muslim societies. Failure to prevent incorporation was influenced, to a large extent, by the intervention of the Western colonial powers: without the interference of the United States in Mindanao and of Great Britain in Patani, the two communities might not have been set upon their present courses of history (see Chapters 1 and 5). Also, the Moro and the Malay societies were characterized by internal disunity. In the case of the Moros, the absence of co-operation and co-ordination among various resistance forces was compounded by the existence of different ethno-cultural groups within the Muslim community. Although there was no ethnic division within Patani society, cleavages between religious and non-religious leaders and within the religious leadership group, especially between traditional and non-traditional religious leaders, contributed to the disunity of the resistance units. Furthermore, some Muslim leaders were willing to collaborate with the government for personal interests.

Another similarity of Muslim resistance in the two countries was that the movements persisted and became transformed. Their persistence was primarily the response to continued integration efforts which threatened their Muslim identity. The extent of this response seems to have corresponded with the degree of repressiveness of government policies. Both the Moro and the Malay resistance movements changed over time, depending on internal and external intervention and stimuli, from unorganized resistance movements to highly organized fronts demanding autonomy or independence, and from organized fronts to qualified participation.

The most outstanding parallel between the two Muslim movements, however, is that Islam became the primary rationale for their resistance to integration. This will become more evident as the contemporary separatist movements of the Moros and the Malays are discussed in the next chapter.

1. Samuel K. Tan's *The Filipino Muslim Armed Struggle, 1900-1972* gives comprehensive accounts on the Moro armed resistance struggles from 1901 to 1941. However, the book has been strongly criticized by George C. Bentley (1978) (and Bentley's review has been equally strongly criticized by Tan (1979)).

2. The preamble to this Act stated that the United States was determined to relinquish sovereignty over the Philippines and to recognize its independence. It also acknowledged the willingness to give Filipinos the greatest possible responsibility in governing their domestic affairs so as to prepare themselves for the control of the independent Philippines.

3. Saleeby (1913: 23) defined the Moro problem as that of finding the 'method or form of administration by which the Moros and other non-Christians who are living among them, can be governed to their best interest and welfare in the most peaceful way possible, and can at the same time be provided with appropriate measures for their gradual advancement in culture and civilization, so that in the course of a reasonable time they can be admitted into the general government of the Philippine islands as qualified members of a republican national organization'.

4. Two factors make a *hajji* respectable in the eyes of the Moro villagers. First, a *hajji* is viewed as a fortunate individual in the sense that God has honoured him to be among those who are able to fulfil all five essential religious duties. Secondly, most *hajji* on their return home are expected to improve their religious and social activities and habits and to become 'better' persons. Such improvements gain respect from the people, though there are *hajji* who are condemned for failure to act as expected.

5. *Ustaz* is an honorary title given to a teacher or *mudarris*. Traditionally, *ustaz* connotes a teacher with an Arabic-educated background who has attended local or foreign Arabic institutes. Presently, most teachers who teach in *madaris* are known as *ustaz*.

6. Kamlon, who led a rebellion against the Philippine government during the first half of the 1950s, was a devout Muslim. He was a *hajji* (a title which is conferred only on those who have performed the *haj*), well versed in Quran reading, and a frequent contributor to the construction of mosques (Tan, 1977: 114). With an estimated 300 followers and local supporters, Kamlon fought government troops from early 1951 to mid-1952. At the end of July 1952, he and 50 of his followers laid down their arms after a series of

peace gestures, but they soon returned to the hills charging bad faith on the part of the government. Four months later, Kamlon and some of his followers surrendered and were sentenced to life imprisonment, but they were immediately granted parole and some 12 000 hectares of land in Tawi-Tawi. In 1953 Kamlon went back to his armed struggle when he was charged with violating his parole. In 1954 the government employed the Jolo Task Force (JOTAF) and the Sulu Air Task Group (SATAG) in an attempt to suppress the Kamlon disturbance. The government operations cost many Moro lives. Kamlon surrendered in 1955 and remained in prison until he was pardoned by President Marcos in 1968 (Tan, 1977: 115-17).

7. There are 54 verses in the Quran on matters concerning *jihad*, and there are at least 283 Hadith regarding the subject related by Al-Bukhari, one of the most authoritative sources on the Prophetic traditions.

3
Muslim Separatism: The Moros and the Malays

THE early manifestation of political and religious antagonisms grew to full-fledged organized separatist fronts as the Philippine and Thai governments pursued their policies of national integration. Muslim leaders in Mindanao and in Patani believed that the Muslim community would be best preserved if there were provisions in the national laws for certain Muslim areas to be given some degree of autonomy. However, it was obvious that the Manila and Bangkok governments were not in favour of such settlement. Some of the Moro and Malay leaders, both secular and religious, took the position that only through separation and independence from the centre would the Muslims be able to preserve their identity, to realize their aspirations, and to develop their rich resources for the benefit of the Muslims.

The Moros

The war between the Moros and the Marcos government started about a month after the declaration of martial law in September 1972 (Noble, 1976: 405). Like other Moro wars in the past four centuries, it was motivated by an endeavour to defend Islam, *bangsa*, and homeland. The war was precipitated by the government's acceleration of integration programmes during the 1950s and 1960s that brought significant numbers of Filipino Christians into Mindanao (see Chapter 2). Among the immediate significant events that led to the formation of organized fronts and to the war of liberation were the Jabidah Massacre (Corregidor Incident) in 1968, the Manili Massacre in 1971, the election of 1971, and the declaration of martial law in 1972.

Details of the Jabidah Massacre are less than clear because of conflicting reports. However, between 28 and 64 Moro recruits out of a large number undergoing guerrilla warfare training in Corregidor Island were massacred in late March 1968 by Philippine Army men. The training was allegedly in secret preparation for Philippine military operations in Sabah—code-named 'Operation Merdeka'. 'Operation Merdeka', as explained by some Moros, was an attempt by Manila to split Islamic ranks and provoke a war between Sulu and Sabah. The cause of the execution was never made public by the Philippine government. According to the lone survivor, Jibin Arola, the 'trainees were shot because they refused to follow the order to attack Sabah' (quoted in Jubair, 1984: 73). Aware of the possible impact of the leakage of this secret plan, the military authorities executed the entire company so that none survived to tell the story (Jubair, 1984; see also Lucman, 1982). The Jabidah Massacre had two important political consequences. First, the Moros were angered at the disregard for their lives shown by the Marcos government. Secondly, it inflamed the Malaysian government of Tunku Abdul Rahman, which, having made compromises to Marcos's desire to establish diplomatic relations, saw itself 'stabbed in the back' by the Philippines (Noble, 1983: 46) which not only pursued Diosdado Macapagal's claim to Sabah, but also trained the Moros to 'invade' it (the Philippine claim to Sabah is discussed in Chapter 5). Two months after the Jabidah incident, Datu Udtog Matalam founded the Muslim Independence Movement (MIM). Its manifesto (Appendix 2) accused the government of pursuing a policy of 'extermination' of the Muslims. The Malaysian government of Tunku Abdul Rahman responded to the incident by promising the Moro leaders, especially Rashid Lucman, that it would help train and provide arms to the dedicated young Moros (Lucman, 1982: 2). In 1969 the first group of young Moros (comprising 67 Maranaos, 8 Maguindanaos, and 15 Suluanos), who were recruited by Rashid Lucman and the MIM, were sent to Malaysia for military training.

From mid-1970 to 1971, violence began to erupt in Lanao del Norte, Cotabato, and Lanao del Sur between Muslims and Christian settlers. The Muslim groups, identified as 'Barracudas' and 'Blackshirts', were allegedly linked with Congressman Ali Dimaporo in Lanao del Norte and with the MIM in Cotabato respectively. The Christian groups, known as 'Ilagas' ('Rats'), were allegedly linked with Governor Arsenio Quibranza of Lanao del Norte and with Ilongo settlers, Tiruray tribal people (see Schlegel, 1979b), and Constabulary units in Cotabato. By the end of 1970, fighting between these rival groups had resulted in many casualties, disruption of the economy, and mass evacuation. More than 30,000 Muslims, Christians, and Tirurays had been forced to leave their farms (McAmis, 1974: 46; Stewart, 1972).

The most publicized incident after the Jabidah Massacre was the Manili Massacre. It occurred in June 1971 when about 65 Muslims—men, women, and children—were murdered by Ilagas at a mosque in Barrio Manili, North Cotabato. To the Muslims, the Manili incident carried special weight because it took place in a mosque compound. It was seen as an act of religious humiliation. As Ali Treki of Libya stated, 'We believe the conflict is now a religious war' (*Philippines Herald*, 8 July 1972). When the late Libyan Information and Foreign Minister, Saleh Bouyasser, was informed about the killing of the Muslims during his visit to the Philippines in 1971, he took the initiative to meet the Moro leaders. It was as a result of this meeting that, on the one hand, Bouyasser recommended to his government that it should help the Moro people, and, on the other hand, a declaration of unity was signed among different groups of Muslim leaders. Cries of 'genocide' began to be heard from Moro leaders.

The hostilities between the Muslims and Christians in the region escalated greatly as the November 1971 election drew near. The number of evacuees on both sides increased to 50,000 (Gowing, 1979. 194). When the election was over, political power in parts of Moroland had shifted from Muslims to Christians. This shift stimulated the two sides to increase their hostilities and thus attracted the attention of overseas Muslim states (Noble, 1976: 410; Gowing, 1979: 195). In 1972, the atmosphere in Mindanao was tense as sporadic clashes between Ilagas and Philippine Armed Forces, on the one hand, and Barracudas and Blackshirts, on the other, occurred here and there.

The imposition of martial law on 21 September 1972, together with the government's attempt to disarm the populace, precipitated the 'war of liberation'. As Noble (1976: 411–12) summarized it:

Marcos' declaration of martial law broadened the base of support and determined the timing of the resort to warfare by the core-group of Muslim radicals. Three characteristics of martial law were critical. First, the centralization of the regime left power almost exclusively in 'Christian' hands: Marcos, his family and associates; 'technocrats' in Manila; and the military. Second, by restricting the range of legitimate political activity the regime left as options only the acceptance of the regime and its promises, or anti-regime revolutionary activities. Third, the regime's immediate moves to collect guns from civilians meant that compliance removed the potential for an eventual resort to force. Thus both Muslims who had been frustrated under the old system but had been able to channel their frustration into nonviolent political activities, and opportunists ready to seize any chance to achieve immediate goals—for power, wealth, or pride—became willing to join the radicals.

The Moro war of liberation began in Marawi City[1] on 24 October 1972 when a force of several hundred men (comprising seven different groups of Maranao youths led by traditional and secular élites) attacked the headquarters of the Philippine Constabulary in the city and seized the Mindanao State University campus (McAmis, 1974: 52–3; Hermosa, 1980: 7–8). The rebels appealed for Muslim support by reasoning that 'Since the Spanish times the government of the Philippines had always been against the Muslims and that it is necessary to overthrow the government so that there would be no restrictions on the practice of Islam' (Gowing, 1979: 196). The clashes between the government forces and the Moro groups lasted 24 hours. The government regained control of the city and the rebels withdrew to the hills. The unrest subsequently spread into rural and urban areas throughout the Moro provinces; and the Moro liberation struggle began.

As the issue of the Muslims in the Philippines captured the attention of leaders in various Muslim countries, the Islamic Directorate of the Philippines (IDP) was organized (after the visit to Mindanao of Libyan Foreign Minister Bouyasser) to serve as a centre for receiving assistance for the Moros. The IDP's organizers were Dr Cesar Majul (Chairman), Macapanton Abbas (Secretary), Senator Domocao Alonto, Senator Mamintal Tamano, Congressman Salipada Pendatun, Congressman Ali

Dimaporo, Congressman Rashid Lucman, Datu Mama Sinsuat, Sult. Amilkadra Abubakar, Mrs Zorayda Tamano, Abdul Karim Sidri, Musi. Buat, Farouk Carpizo, and Nur Misuari. They signed a declaration of unity which 'declared the readiness of the Muslims to defend Islam, the Homeland, and their people against all forms of agressions against the *Ummah*' (MNLF, 1982: 7).

The Moro Liberation Fronts: MNLF, BMLO, MILF, and MORO

There exist five Moro underground groups currently active in the southern Philippines: the Misuari and Pundato factions of the Moro National Liberation Front (MNLF); the Bangsa Moro Liberation Organization (BMLO), recently renamed Bangsa Muslimin Islamic Liberation Organization (BMILO); the Moro Islamic Liberation Front (MILF); and the Moro Revolutionary Organization (MORO). In order to understand the essence of the Moro struggle, it is essential to investigate how the fronts have emerged.

The Emergence of the MNLF

The uncertainty and fears generated in the aftermath of the Jabidah Massacre in 1968 created a new urge among the Moros to search for alternatives to secure the *ummah*. To the Moros, the Filipino Christian government had proved to be insensitive to their demands and unwilling to ensure protection of their lives (Asani, 1985: 306). Amid this state of uncertainty many organizations emerged, including the MIM, Ansar El-Islam, and Union of Islamic Forces and Organizations (UIFO). The MIM aimed at bringing together the Moro provinces as an independent Muslim state. It had the effect of raising the Moro struggle to a more advanced level. But its strength was limited because the organization revolved around only a few aristocratic political leaders. The MIM disintegrated when Datu Udtog Matalam surrendered to the government in December 1972.

This period also saw Muslim youths and student activists beginning to assert their demand for better treatment of the Moro people. For example, Nur Misuari, who later became Chairman of the MNLF, was an officer of the Kabataan Makabayan (Patriotic Youth), a radical student group in Manila which applied a Marxist analysis to the Philippine situation and advocated a revolutionary struggle against feudalism, capitalism, and imperialism (Shoesmith, 1983). Macapanton Abbas, who later served as Secretary of the BMLO, was leader of the National Union of Students of the Philippines (NUSP) from 1967 to 1969. Salamat Hashim, who became Chairman of the MILF, was one of the active leaders of the Philippine Students' Union in Cairo; he returned to the Philippines in 1970 with the intention of working to restore Islam among the Muslims (Glang, 1976: 7; Jubair, 1984: 77–8).

The MNLF was formed in 1969 by a group of young secular-educated Moros who were among the recruits sent to Malaysia for military training. During their training, a seven-member committee of the Front, with Nur Misuari as Chairman and Abul Khayr Alonto as Vice-Chairman, was organized without the knowledge of Rashid Lucman and other leaders who had recruited them. This was because the younger generation of Moro leaders like Nur Misuari, Otto Salahuddin, and Ali Alibon wanted to dissociate the MNLF from the traditional aristocratic élite whose leadership was viewed as feudal and as 'a party to the oppression being waged against the Bangsa Moro people' (George, 1980: 201–2).

At the convention of the Moro Youth National Assembly in Zamboanga City, Nur Misuari and other younger-generation leaders agreed to a resolution that the Moros would strive for 'a federal state' and they stood as candidates in the elections for the delegates to the constitutional convention of 1971. This suggests that the MNLF leaders were at this stage not ready to expose their own Front. Instead, they attempted to pursue their struggle through legal channels and to test their strength against the traditional leaders by contesting the elections. They failed to win the contest.

Having made the arrangements for the foreign training, Congressman Rashid Lucman organized the BMLO when the first batch of 90 trained guerrillas returned to Moroland in 1970 (BMLO, 1978: 1). The BMLO considered itself to be an umbrella organization of all the liberation forces. Rashid Lucman was head of the Supreme Executive Council, Macapanton Abbas Secretary, Udtog Matalam Jr. head of Military Committee for Kotawato, Abul Khayr Alonto for Ranao, and Nur Misuari for Sulu (MNLF, 1982: 7). Misuari and Abul Khayr Alonto used the BMLO as a foundation in strengthening the MNLF.

Unity of the Moro leaders seemed to have been achieved, at least temporarily, after the Islamic Directorate of the Philippines was formed in 1971. Rashid Lucman, Senator Domocao Alonto, and Senator Salipada Pendatun went to Libya as representatives of the Moro people and met Muammar Al-Qadhafi to secure Libya's assistance to the Moros. Qadhafi promised them that he would provide 'all forms of assistance' to the liberation movement (MNLF, 1982).

After martial law was declared in September 1972, Macapanton Abbas went to Jeddah to present the Moro case to the Organization of the Islamic Conference (OIC). He submitted a 200-page report on the Moro struggle to Tunku Abdul Rahman, the OIC Secretary-General. At about the same time, Misuari went to Libya to follow up the promises of the Libyan government. He took this trip as an opportunity to introduce the MNLF to the Moros and Muslims abroad. Soon after Nur Misuari returned to Libya, Salamat Hashim joined him and other MNLF leaders. Together, they were able to convince the Libyan officials, who came to Sabah to deliver financial assistance to the Moro movement, that the assistance should be given to the MNLF. How they were able to do so is not clear. One can only speculate that perhaps a revolutionary government like that of Libya would rather see young energetic leaders like Misuari and his group, as opposed to the old traditional élite and politicians, lead a revolutionary struggle. Moreover, the Libyans would not want their funds to be used ineffectively by some politicians whose reputation for probity was poor. Besides, the MNLF was ideologically more in tune with Libya. The BMLO leaders, particularly Rashid Lucman and Macapanton Abbas, however, accused Misuari's group of 'betrayal' and 'counter-revolution' (MNLF, 1982: 7; Lucman, 1982: 4–6). In Lucman's words (1982: 4; BMLO, 1978: 1):

> I was informed later by Minister Saleh Bouyasser, while in Tripoli, that the letter sent to me was sent upon the instructions of President Qadhafi to meet Treki immediately in Sabah because he was bringing with him the money I requested. That money was amounting to US$3.5 millions. I was not able to meet Treki because Salamat translated the letter written in Arabic and upon knowing the contents, they, Misuari, Salamat, and some Sulu pseudo-leaders went to Sabah to meet Treki, informing Treki that I was not able to come because the Philippine government did not allow me to go out of the country but instead authorised the group to meet him, which was a lie.

Be that as it may, the rift between the younger-generation and traditional aristocratic leadership groups finally came into the open. Nur Misuari and his group moved to Sabah and then to Tripoli and officially formed the MNLF Central Committee (see Table 4.7). The Front issued its manifesto (Appendix 3) on 28 April 1974. With financial assistance and arms supplied by Libya and Sabah, the MNLF began to take overall charge of the armed conflict in Moroland by providing weapons and other supplies to the Muslim groups which were already at war with government forces and to those who wanted to join the struggle. At the same time, it exerted its efforts to gain recognition from the OIC. In a way, the MNLF was viewed by repressed Muslims, who needed a front as instrument to fight the encroachment of the Filipino Christians, as a vehicle of *jihad*.

While the MNLF carried out its armed struggle, the BMLO leaders—Rashid Lucman, Macapanton Abbas, and others—who failed to secure assistance from Libya, agreed to 'co-operate' with the Marcos government. They argued that their purpose was 'to inject the rationale of the Moro struggle into government policies in order to lay the basis of the legitimacy of the Bangsa Moro Struggle' (MNLF, 1982: 9). For example, in 1973 Macapanton Abbas, Gibril Ridha, Napis Bidin, and others joined the Presidential Task Force for the Reconstruction and Development of Mindanao (PTF-RDM), created, among other things, to restore peace and order in the Moro region and to implement selective amnesty and rehabilitation (MNLF, 1982: 10; Mastura, 1984: 248). In an attempt to promote further divisions within the Moro leadership, Marcos acknowledged Rashid Lucman as the 'Paramount Sultan of Mindanao and Sulu' in May 1974. In June of the same year, Sultan Rashid Lucman and other Moro leaders organized a conference on 'Government Policies and Programs for Muslim Mindanao' which was financed by the government (MNLF, 1982: 11; Ansar El-Islam, 1974: 1).

The conference, which was held at Mindanao State University on 4–6 June 1974, adopted a resolution demanding autonomy. More significantly, the document of this resolution, which was signed by 20,000

Moros, was attached to the report of the Quadripartite Ministerial Commission (QMC)[2] to the meeting of the Islamic Conference of Foreign Ministers (ICFM) in Kuala Lumpur in the same month. It was this QMC report that became the basis for the Kuala Lumpur Resolution No. 18 (see Appendix 9) urging the Philippine government to undertake a political and peaceful solution to the Moro problem through negotiations with the Moro leaders, particularly the MNLF (ICFM, 1974). Since the conference was designed by Manila to draw support from Moro leaders, particularly the traditional élite and Muslim government officials, its resolution demanding autonomy disappointed the government. As a consequence, Marcos labelled Sultan Rashid Lucman and some other Muslim leaders as his opponents (MNLF, 1982: 11). In 1975 Sultan Rashid Lucman, Salipada Pendatun, and Macapanton Abbas left the Philippines for Saudi Arabia, where they attempted without success to unite with the MNLF. Abbas left behind him the PTF-RDM, which was eventually incorporated into the Southern Philippines Development Administration (see Chapter 5).

At the Sixth Islamic Conference of Foreign Ministers held in Jeddah in July 1975, the MNLF under the leadership of Nur Misuari was given formal recognition by the OIC, despite the fact that Sultan Rashid Lucman and Macapanton Abbas were the ones who originally submitted the Moro case to the OIC under Tunku Abdul Rahman. The OIC was convinced that the MNLF was dominating the leadership of the Moro struggle. Resolution No. 18 of the Fifth Islamic Conference of Foreign Ministers in Kuala Lumpur in 1974 had urged the Philippine government to negotiate with representatives of the MNLF. Libya, which played an important role in the Moro movement, supported the MNLF, and Marcos sent a delegation headed by his Executive Secretary, Alejandro Melchor, to Jeddah in January 1975 to negotiate with the MNLF leaders.

The MNLF was able to convert sporadic clashes between the Moros and the Marcos regime to a 'conventional war' which became the Philippines' most serious internal conflict since the Communist-led Hukbalahap Rebellion of the late 1940s. The ability of the MNLF to escalate the war during 1973 and 1976 and to stalemate it was considered a significant success, though the toll of the war was tremendous.[3] The MNLF gained international attention, particularly from member countries of the OIC, and forced the Philippine government to concede by signing the Tripoli Agreement (see Chapter 5). The rapid ascendancy of the MNLF, however, can be attributed not so much to effective organization as to a fortuitous combination of factors. Prior to the MNLF's take-over of the leadership of the struggle, the Moro resistance against the Marcos regime had been carried on by various independent groups, especially in Lanao, Cotabato, and Sulu. The MNLF's principal contribution was to consolidate these existing forces. They included not only armed guerrilla units, but also most of the villagers in the war-affected areas who seemed to be directly or indirectly involved in the struggle. The response of some Muslim countries to the plight of Muslims in

Mindanao was also a major factor contributing to the Front's success.

But the MNLF's success was short-lived. Hopes for the implementation of the Tripoli Agreement were shattered. President Marcos insisted on holding a plebiscite on the essential provisions of the Agreement itself. When the plebiscite produced the result he expected, Marcos 'implemented' the Agreement as he chose (Noble, 1984: 9). The ceasefire collapsed and fighting resumed in late 1977. In March 1980, a spokesman of the Philippine Foreign Ministry issued a declaration renouncing the Tripoli Agreement as 'null and void' (Misuari, 1981: 2–3; 1983a: 3), though the government later denied this renunciation. In response, the MNLF officially announced the assertion of its original position of self-determination and independence at the Third Summit of the Islamic Heads of States in Mecca in January 1981 (Misuari, 1981: 4). The struggle of the MNLF continued.

Organization

The uprising of an unorganized mass determined to change the immediate circumstances of their lives is a frequent event in the initial stages of revolutionary movements. A study of revolutionary motivation suggests that spontaneous forces of peasants, workers, students, or minorities can be suddenly galvanized into collective action by a wide variety of incidents and circumstances. The spontaneous actions of the mass are not necessarily aimed at revolutionary change, but may be aimed only at the redress of immediate grievances. It is the function of revolutionary organization, leadership, ideology, and subsequent events to provide the movement with a more general purpose and revolutionary direction (Greene, 1974: 60–2). The role of organization, then, is essential to the realization of revolutionary goals. Spontaneity may indicate the potential power of revolutionary sentiment, but only organization can give it lasting consequence.

From a conventional standpoint, the strength of a revolutionary organization is measured by the extent to which it institutionalizes the recruitment and training of its cadres, establishing mechanisms for their continuing indoctrination in the norms of the movement, maintains channels of communication that facilitate feedback from the rank and file to the leadership, and constructs functionally specific suborganizations charged with clearly defined tasks. A 'strong' revolutionary organization maintains a clearly established chain of command and has a minimum of factional conflict. And a revolutionary organization is likely to be strong if its ideology is a coherent and highly articulated system of values connected to prescriptions for revolutionary action. History is filled with instances of unsuccessful revolutions that failed to meet these criteria for revolutionary organization (Greene, 1974: 65).

In the case of the Moro liberation organization, some of these criteria are absent. The MNLF, for instance, did not experience the normal process of organizational development by which the machinery of organization becomes institutionalized. Instead, the Front was created instantly as a region-wide network of organizations in an attempt to consolidate the various existing Moro forces that were fighting against the Marcos regime. Such organization was not possible without the outside assistance of Muslim leaders such as Qadhafi of Libya and Tun Datu Mustapha Harun of Sabah. This external assistance contributed to the Front's rapid ascendancy as well as to a dependence upon foreign support.

It is unrealistic to suggest that the MNLF was a well-structured organization, with a clearly established chain of command running from the top of the organization's hierarchy to the bottom, comparable to that of the National Liberation Front of South Vietnam (NLF) (cf. Pike, 1966). Even during its most active period, from early 1973 to late 1977, the effective structure of the MNLF was not as it was constituted on paper (Appendix 4).

The basic components of the MNLF organizational structure consisted of a Central Committee, headed by a chairman chosen from among its members, and provincial committees. The Central Committee, officially formed in Libya in 1974, functioned as the 'executive' body of the Front. It comprised about 13 members (MNLF, 1982: 12) and different bureaus or committees which performed various duties (cf. Noble, 1976: 412; cf. Hatimil Hassan, 1981: 253–6). The identity of the Central Committee members was not generally known, except for certain senior leaders (see Table 4.7). This was due to the nature of the underground movement, to the constant change of individual members, and to the fact that the Committee was run by a few 'strong' leaders. Unlike most conventional revolutionary organizations, the MNLF Central Committee operated outside the Moro homeland. Its leaders directed their efforts mainly at cultivating external support, leaving the task of fighting and organizing to local leaders. It was largely the achievement of the Central Committee that the Front continued to receive support from certain Muslim countries and to be recognized by the OIC.

In the homeland, provincial revolutionary committees were established in different Moro provinces. They were divided into three groups: Sulu, Kotawato, and Ranao. The grouping coincided with the three major ethno-cultural groups: Suluanos, Maguindanaos, and Maranaos. Three strong leaders of the Front—Nur Misuari, Salamat Hashim, and Abul Khayr Alonto (later Dimas Pundato)—also came from the three respective ethnic groups. The major functions of the provincial committees were to consolidate the existing Moro armed groups within the province, to recruit and train Moro fighters, and to fight the war. But the boundaries differentiating Moro groups fighting under and outside the Front remained fluid (Noble, 1976: 412).

Members of provincial and barrio committees and of armed units of various sizes of the Bangsa Moro Army (BMA) were regarded as members of the Front. The members covered by such groups were estimated to have ranged from 15,000 to 30,000. In addition, more than 50 per cent of the Moro population were claimed by the Fronts to support their cause (Murad, 1982; Pundato, 1984). However, if one defines a member as one who signs an undertaking to the organization and thereafter regularly pays his subscription (cf. Duverger, 1976: 62–90), then the MNLF had no members. It had no formal system of registration or of any regular collection of subscription, except alms provided by villagers. On the other hand, Muslim villagers were viewed by the Front as its members because they were potential active participants of the movement. Despite the fact that members of provincial committees and commanders of different armed units were men of different backgrounds, the majority of them were drawn from the leadership group that was dominant in their respective areas. Thus, in Lanao, committee members and unit commanders comprised largely traditional aristocrats whereas in Sulu they were commoners. In other words, the movement was led and supported by the local élite.

The MNLF was a loosely knit organization. It did not manage to bring into its structure all the Moro groups fighting the Marcos government or to control the behaviour of many groups under it (Noble, 1976: 415). The Front was unable to construct a clearly established chain of command. 'One of the flaws in the MNLF leadership structure is that there seems to be a wide gap in communication between the Central Committee and the Field Commanders,' writes Madale (1984: 181). Each provincial committee acted on its own initiative. Links among them were also weak, though they occasionally reinforced each other in battle. The Central Committee contented itself with setting broad policy outlines and organizing external support, leaving local leaders to make their own decisions.

One of the difficult tasks confronting the movement was logistics. Its supply depot was in Sabah, where weapons, military equipment, and other supplies from foreign supporters were received. From Sabah, weapons, ranging from carbines and rifles to machine-guns and mortars, and other defence equipment were shipped to Sulu and Mindanao by sea. Some of the shipments failed to reach their specific destinations, and Nur Misuari was accused of distributing supplies in favour of Suluanos (Madale, 1984: 181; BMLO, 1978: 1; MNLF, 1982: 9). The difficulty, however, lay in the fact that adequate and efficient transportation and communication systems were very difficult to organize. Those who did not receive adequate supplies of arms from the MNLF sometimes managed to gain access to weapons from some government troops whose morale was low.

At the height of its influence, the MNLF's financial support came mainly from external sources. The government of Libya provided the largest share (Lucman, 1982: 3–5; BMLO, 1978: 1) while the Sabah state government of Tun Datu Mustapha Harun allegedly allowed the Moro fighters to be trained in Sabah and helped the Front in soliciting and buying weapons. Another major external source of finance was the Islamic Solidarity Fund (interview with officials of the Organization of the Islamic Conference, Jeddah, 25 and 26 May 1986; Pundato, 1984: 1–5).[4] In addition, some government agencies, foundations, and companies in Muslim countries contributed funds to the MNLF in the form of *zakat* (alms). Such support is legitimate and sanctioned by Islam; for it is obligatory for every Muslim to give *zakat* to, among others, those who struggle in the name of Islam. Internal financial support to the movement was also mainly in the form of *zakat*. Many villagers provided their alms to the movement in kind, such as rice, crops, and animals.

The MNLF placed the responsibility for recruitment on the shoulders of the provincial revolutionary committees. Recruitment into fighting units often involved the manipulation of religious and nationalist sentiments. Potential recruits were reminded of their religious duty to participate in the *jihad*. As the Quran (IX: 41) urges, 'Go ye forth (whether equipped) lightly or heavily, and strive and struggle, with your goods and your persons, in the Cause of God. That is best for you, if ye (but) knew.' They were also normally advised of their *maratabat* (honour) by a warning that 'a man who has lost his *bangsa* (identification with ancestors) has no *maratabat*'; and 'a man without *maratabat* is nobody' (Saber, Tamano, and Warriner, 1974: 233). Since the war was between two clearly defined groups, Muslims versus Christians, the screening process of recruitment was less complicated. Many young Muslim villagers who were willing to fight were given arms and basic military training. Some villages in the remote areas were converted into training camps. Many fighters and commanders were trained in Sabah. Selected individuals were also sent to Libya, Syria, Egypt, and training camps of the Palestine Liberation Organization (PLO). Recruitment was a relatively open and easy task. Many Muslims had been forced into fighting by the war itself (Noble, 1976: 414), and the Moros in general were strong in religious and nationalist sentiments. Indeed, the MNLF enjoyed popular support. Mass mobilization was usually undertaken not only by leaders of the Front, but also by different leadership groups in the Moro communities. The strength of the movement, therefore, does not lie in the quantity of its armed recruits alone: the potential mass support when circumstances arise must also be taken into account.

Fragmentation of the Movement

The breakdown of the Tripoli Agreement shook the credibility of the MNLF leadership, allowing the cleavage which had always existed within the Front to emerge (Noble, 1978: 3–4). Salamat Hashim's group was first to break away from the mainstream MNLF. The split began when Salamat Hashim, a member of the Central Committee and chairman of the Committee for Foreign Affairs, attempted to take over the chairmanship of the MNLF by challenging Nur Misuari at a meeting in Mecca in December 1977. The attempt was supported by Sultan Rashid Lucman of the BMLO, Domocao Alonto of Ansar El-Islam, and Salipada Pendatun of the Muslim Association of the Philippines (MAP). Nur Misuari refused to recognize the election which resulted in favour of Salamat Hashim as Chairman; he discredited Hashim as incompetent and insubordinate and accused him of treachery to the MNLF (Noble,

1978: 5; George, 1980: 263). The MNLF leadership issue also became entangled in a wider intra-Arab dispute. While Egypt was in favour of Salamat Hashim, Qadhafi described him as an agent of President Anwar Sadat. In his letter to the Secretary-General of the OIC, Dr Ahmadu Karim Gaye, Hashim (1977: 2) gave the following reasons for the take-over:

Mr Misuari, who assumed the MNLF Central Committee Chairmanship, plunged the MNLF in a deep leadership crisis, due to multifarious reasons among which are: '1. The MNLF was being manipulated away from Islamic basis, methodologies and objectives and fast evolving towards Marxist–Maoist orientations. 2. Instead of evolving towards harmonized, unified and collective leadership the Central Committee has evolved into a mysterious, exclusive, secretive and monolithic body, whose policies, plans, and decisions and dispositions—political, financial and/or strategic—became an exclusive preserve of Nur Misuari.... 3. This mysterious, exclusive, and arrogant nature of the MNLF leadership resulted into [sic] confusion, suspicion and disappointments among the members of *mujahideen* in the field resulting in the loss to the cause of a great number of fighters'.

In March 1984, Hashim's faction became a separate organization called the Moro Islamic Liberation Front (personal communication with Muammar Qutb, Head of the MILF in Mecca, Saudi Arabia, 29 March 1984). It saw no hope for future control of the MNLF leadership or for reconciliation with the Nur Misuari Faction. In addition, Hashim, a religious-educated aristocrat, wanted to emphasize the Islamic orientation of his group by detaching it from the MNLF which he regarded as a left-leaning organization.

In March 1978, Abul Khayr Alonto, Vice-Chairman of the MNLF, and his followers surrendered to the government. One of the reasons given was that Nur Misuari had removed him from the MNLF Central Committee because he was not in favour of Misuari's manoeuvring for complete independence (George, 1980: 263; NMRC-MNLF, 1978: 1–8). Perhaps Abul Khayr Alonto, being an aristocrat, also disagreed with Nur Misuari's persisting view on the need for revolutionary change of Moro traditional society. The exit of Salamat Hashim and Abul Khayr Alonto gave the opportunity for Dimas Pundato, Chairman of Ranao Revolutionary Committee, to be promoted to Vice-Chairman of the MNLF.

In March 1982, however, Dimas Pundato announced the formation of the MNLF-Reformist Group. Before his announcement, Dimas Pundato and some 44 others submitted a nine-point reform proposal (Appendix 5) to Chairman Nur Misuari. The proposal was rejected, and Dimas Pundato and his men were dismissed from the Front (interview with Commander Iqra Sheikh Mokhtar, Chairman, Ranao Sur Revolutionary Committee, Lanao del Sur, 17 December 1985). In June of the same year, the Reformist Group met in Tawi-Tawi and adopted resolutions rejecting Nur Misuari's leadership and calling for autonomy instead of independence (Appendix 6).

The BMLO re-emerged following the breakdown of the Tripoli Agreement. Sultan Rashid Lucman saw the possibility of forging unity among the different Moro forces. In December 1977 the BMLO signed the 'Declaration of Principles and Covenant of Unity' with the Salamat Hashim Faction, Ansar El-Islam, and the MAP (BMLO, 1978: 1; Lucman, 1982: 7). This lasted until April 1978 when Salamat Hashim failed to join Sultan Rashid Lucman and Salipada Pendatun (the leader of the MAP) in an attempt to initiate unity talks with Nur Misuari's Faction during the Ninth Islamic Conference of Foreign Ministers in Dakar, Senegal (Lucman, 1982: 6; George, 1980: 262). In 1979 an effort, assisted by the Muslim World League (MWL), to establish a united front between the BMLO and Nur Misuari's and Salamat Hashim's groups also failed. The most publicized unity attempt was made under the auspices of the Muslim World Congress (MWC) in January 1983 in Karachi, Pakistan. The effort again failed because the so-called 'Peace and Unity Dialogue of Filipino Muslims' was dominated by traditional élite, Muslim officials of the Autonomous Regions, and a number of surrenderees (MWC, 1983). The leaders of the three major factions—Nur Misuari, Salamat Hashim, and Dimas Pundato—boycotted the meeting. In his telegram to Habib Chatti, the OIC Secretary-General, Nur Misuari (1983b: 1) wrote: 'This meeting if pushed through will undermine Islamic Conference's peace mediations to solve the problem of our people.... MNLF therefore condemns this undue interference in our internal affairs most vigorously.' The Karachi meeting was the last unity attempt attended by Sultan Rashid Lucman and Salipada Pendatun. They died in 1984 and 1985 respectively.

Why did these unity efforts among different groups of Moro forces fail? Basically, every attempt at unity has been seen as benefiting one group more than the others. Unity has been attempted among groups of unequal strength and the strongest group has seen no advantage in uniting on an equal basis with weaker groups. Disunity within the Muslim world has also contributed to the difficulty of uniting the Moros.

Disunity or factionalism, however, is a source of strength as well as weakness. While factionalism weakens the struggle in many ways, it also provides a spirit of competition among the factions. Each group seems to try hard to strengthen itself by concentrating on building camps and organizing recruits. Although military activities have been reduced to sporadic clashes, each major faction should be better prepared if 'war' should return to Sulu, Cotabato, and Lanao.

If ideology is defined as 'a blueprint for an alternative social system, a goal for which men can risk their lives in violent combat' (Enloe, 1973: 232), then Islam must be the only foundation of Muslims' ideology. For there is no social system other than Islam for which Muslims are allowed to die. All Moro liberation forces, except the MORO, are founded on a similar basis of ideology—Islam. Furthermore, Islam has become the national identity of the Moros; it is the most powerful symbolic means of mobilizing support and of legitimizing sanctions for revolutionary action. But despite their similar ideological foundation, the Fronts bear their own distinct characteristics. The Nur Misuari-led MNLF, for instance, is known as a 'left-leaning' nationalist faction because of its uncompromising commitment to revolutionary armed struggle against Manila domination of Moroland; because some of its leaders have advocated egalitarianism and because of Nur Misuari's early links with Kabataan Makabayan; and because the group has chosen to maintain close ties with Muslim revolutionary governments such as Libya, Syria, Iran, and Algeria. Leaders and supporters of Nur Misuari's group are largely non-aristocratic nationalists and young activists.

The Manifesto of 1974 (Appendix 3) commits the MNLF to a war of national liberation aimed at establishing 'an independent *Bangsa Moro Republik*' and to 'a democratic system of government which shall not allow or tolerate any form of exploitation and oppression of any human being by another or of one nation by another'. In the words of Nur Misuari (1984: 6–7),

The armed struggle of the MNLF is a revolution for national salvation and human justice.... Like the struggle of our forefathers, our armed struggle today is a *jihad* for national salvation from colonialism, except that while they fought off external aggression, today we are fighting against colonial injustices, exploitation and tyranny imposed from within.... And it goes without saying that the MNLF and the Bangsamoro Revolution are fighting for justice, as our cause is just and legitimate.... In this sense, therefore, the armed struggle of the Bangsamoro people is an instance of *jihad*. *Jihad* is the path of struggle of Muslims, either in the moral, ethical, spiritual or political realm, to bring about a positive transformation of the inner self and the socio-economic and political order.

Nur Misuari's aim of transformation of Moro society as an ultimate objective beyond winning independence, however, brought the MNLF into direct conflict not only with the Muslim aristocratic élite but also with commoners whose deep-rooted communal loyalties were held as important to their self-identity (Shoesmith, 1983: 5). Critics suggest that the Front has been allied with the Communist Party of the Philippines (CPP) and the New People's Army (NPA) since 1975. However, the author's interviews in 1985 with Commanders Solitario Solaiman and Narrah Abdul Jalil, heads of Ranao Norte and of a non-partisan group in Lanao del Sur, respectively, and with Commander Gadil Ismael, aide-de-camp to Nur Misuari, have confirmed otherwise. Solitario told the author that one of the reasons for which Commander Nur Khan, Chairman of Davao Revolutionary Committee, surrendered in 1985 was that he was unable to withstand the pressures from both the government forces and the communist insurgents. However, Commander 'Akilha', Narrah's deputy, suggested that local unit commanders of the MNLF and NPA, whose units happened to be operating in the areas adjacent to one another, normally sought a mutual sanctuary agreement, allowing them to retreat into each other's territory when pursued by government forces. Such arrangements, however, are regarded as unofficial individual understandings. The MNLF has hesitated to associate itself with the CPP/NPA. Nur Misuari has stated: 'We do not have any formal contact with them because they are preoccupied with their own war in the North and the Visayas and we are engaged in our war in the South (*Mahardika*, August 1980). Recently, Nur Misuari has been quoted as saying that he would not hesitate to join forces with the New Armed Forces of the Philippines (NAFP) ('new' to differentiate it from the old Armed Forces of the Philippines (AFP) under the Marcos regime) against the NPA groups (*New Sunday Times*, 16 November 1986). The MNLF leadership believes that communism is incompatible with Islam, and that association with the communists would weaken its relationship with the Muslim world (Noble, 1984: 11).

The MILF, MNLF-Reformist Group, and BMLO may be characterized as conservative. The MILF leadership is derived largely from traditional aristocratic and religious élites who are concerned with the promotion of Islam and the preservation of traditional Moro society. In his handbook, *The Bangsamoro Mujahid*, Salamat Hashim (1985: 51) declares that 'the ideology of the MILF is *La illaha illa Allah Muhammad rasul Allah*' ('There is no god but Allah, and Muhammad is the messenger of Allah'). This illustrates that the MILF gives priority to promoting Islam. Salamat Hashim (1985: 52) writes, 'The core of our political and social program is the Islamization of the Bangsamoro people.' The MILF views the objectives of non-Islamic or Marxist-oriented groups as alien to Islam and refuses to co-operate with them (Hashim, 1985: 26). The Front commands loyalty from its leaders' followers and their clans. It also has support from conservative Muslim countries such as Saudi Arabia, Egypt, Pakistan, Malaysia, and Kuwait. The nature of support given includes moral, financial, and place of sanctuary.

The MNLF-Reformist Group, like the MILF, employs the Quran and Hadith as its basic source of reference. Its ideological orientation and programmes are executed within the guide-lines of Islamic teachings. It aims at establishing an Islamic society within the Moro provinces by means of gradual implementation of the *sharia*. The main effort is to activate political negotiation with the Philippine government on the basis of the Tripoli Agreement. If this effort fails to bring the government to the negotiating table, it would step up military activities in order to put pressure on the government and to draw the attention of the Muslim nations. However, the Reformist Group differs from the MILF in that the former is dominated by traditional and secular-educated leaders, comprising mainly Maranao aristocrats, while the latter is dominated by traditional and religious élites, consisting of Maguindanao aristocrats. This is due partly to the educational backgrounds and personal preferences of the top leaders of each group. For example, the MILF prefers a leader to be well versed in Islamic jurisprudence and have a good understanding of Islam (Hashim, 1985: 38). The Reformists' main international contacts include Malaysia and Saudi Arabia.

The BMILO (BMLO), under the chairmanship of Dr Yusoph Lucman,[5] has as its objective to make Islam supreme within the Philippine state. As Yusoph Lucman explains, 'our objective is not the establishment of an autonomous or independent state or any other worldly gains', but to return to the *jihad* and to work for the restoration of Islam

(BMILO, 1984a: 1; *Arab News*, 9 November 1984). Like the MILF, it stresses the promotion of Islam. Unlike the MILF, the BMILO does not emphasize armed organization. Its leaders also reject 'Moro' as the identity of the Muslims in the southern Philippines. Their main argument is that 'the name "Moros" was given to the Muslims by the enemies of Islam, the Spanish Christian invaders with the objective of insulting the MINSUPALA Muslims' (BMILO, 1984b: 1). They prefer to be known as 'Bangsa Muslimin of MINSUPALA' (interview with Jamil Yahya, Secretary of the BMILO, Mecca, 17 April 1986).

The three-way factional split between Nur Misuari, Salamat Hashim, and Dimas Pundato also reflects lines of ethno-cultural background and personal loyalty. The split coincides with the three major ethno-cultural divisions. The Maguindanaos and the Maranaos, who incline towards preservation of the Moro traditional system, are supporters of Salamat Hashim and Dimas Pundato respectively. The Suluanos, many of whom advocate egalitarianism, are behind Nur Misuari. The ethno-cultural boundaries, however, are not rigid. Crossing ethnic boundary lines is not uncommon. It causes or is reinforced by personal loyalties or ideological orientations. Personal loyalties seem to have been dominant; leaders and followers have fought and surrendered together. Only the leaders are really conscious of adhering to an ideology; the followers are conscious only of being Muslims and feeling discriminated against as Muslims (Noble, 1978: 4).

The Moro Revolutionary Organization was founded in 1982 by young Moro activists and students. It is supported by the CPP and under the direction of its political arm, the National Democratic Front (NDF). Its manifesto (Appendix 7) calls for armed revolutionary struggle, though it has yet to organize its own armed units. MORO appears to have been created as a communist attempt to penetrate the Moro communities.

After the Marcos regime was replaced through a four-day bloodless revolution in February 1986, the new government under President Corazon Aquino tried to negotiate an end to the protracted Moro war. In September 1986 President Aquino met the MNLF leaders and further talks were held in Jeddah in January 1987 between Aquilino Pimentel, a member of the Aquino cabinet, and the MNLF-Nur Misuari Faction. The Jeddah sessions produced no pact, but the two parties agreed to start formal negotiations when Nur Misuari dropped his long-standing demand for independence.

Negotiations for full autonomy of the five southern islands (Mindanao, Basilan, the Sulu group, Tawi-Tawi, and Palawan) as demanded by the MNLF will meet many problems. For instance, autonomy for the region will be 'subject to democratic processes', which means that it must win the consent of Mindanao's Christians, who constitute about 77 per cent of the population of the region. The autonomy demanded by the MNLF may have serious legal implications because the new constitution ratified in February 1987 has its own provisions for the creation of autonomous regions in Muslim Mindanao and the Cordilleras. Furthermore, the Jeddah negotiations failed to include the MILF, the rival faction of the MNLF. As Salamat Hashim, Chairman of the MILF, declared, 'The Jeddah meeting does not have any binding effect on us' (*Asiaweek*, 18 January 1987). Just a week after the Jeddah meeting, the MILF launched a three-day attack against the government, resulting in at least 30 killed and 57 wounded (*Canberra Times*, 17 January 1987). Despite these obstacles, the chief government negotiator, Aquilino Pimentel, stated, 'I believe it can truly be said that this conference is a momentous step towards the formulation of a just and lasting peace' (*Canberra Times*, 10 February 1987). Similarly, the chief MNLF envoy, Habib Hashim, reflected, 'The path to peace is long, arduous and thorny, but it is a path that we have willingly chosen' (*Canberra Times*, 10 February 1987).

Field Studies

Between October 1984 and May 1985 the author visited several camps and 'liberated' villages controlled by the MNLF-Nur Misuari Faction, MNLF-Reformist Group, and MILF. In order to provide some substance to the more general accounts of the factions described above, the following paragraphs report on some of these visits.

The MNLF-Reformist Group, Lanao del Sur

In mid-November 1984 I spent eight days with a guerrilla unit of the Ranao Sur Revolutionary Committee (RSRC) of the MNLF-Reformist Group. Perhaps because I am a Malay-Muslim, the Maranao *mujahideen* accepted me like a brother.

The RSRC was headed by Commander Iqra, a Maranao aristocrat with an MA degree in International Relations from a university in the Philippines. I travelled with Commander Iqra and a group of 15 armed guerrillas to several 'liberated' villages around Lake Lanao. In some villages guerrillas and villagers working in their fields, growing a variety of crops; to the eyes of outsiders, guerrillas and villagers are indistinguishable. In every village we visited, the villagers were very co-operative and hospitable.

We crossed Lake Lanao by motor boat to a village where more guerrillas and commanders from neighbouring villages joined us. In this village, I spent four days living with a group of about 30 guerrillas and commanders. The guerrillas were typical young Moro villagers who were trained as fighters. I was informed that most fighters were allowed to return to their respective villages as reserves after completion of several months' training. They can be called to active duty when required. 'In the case of emergency,' remarked Commander Iqra during our conversation, 'we can easily mobilize 10,000 fighting men.' Many of the commanders had left school to join the struggle between 1972 and 1976. Some were recruited and trained in Libya and Malaysia. The guerrillas and commanders were mostly equipped with light conventional weapons including M14, M16A1, M1819A1, and Russian AK 47 rifles, M1 carbines, M1911A1 and other types of pistol, M79 grenade-

TABLE 3.1
Potential Muslim Supporters Claimed by MNLF-Reformist Group, 1984

Province	Potential Muslim Supporters (per cent)[1]
Basilan	40
Davao del Sur	45
Lanao del Norte	50
Lanao del Sur	90
North Cotabato	20
South Cotabato	65
Sulu	20
Tawi-Tawi	90
Zamboanga del Norte	95
Zamboanga del Sur	35
Average	55

Source: Pundato (1984: 4).
[1]The percentage distributions are based on the total potential Muslim supporters in each province. They form about 60 per cent of the total Muslim population, considering 15 per cent are with the government and 25 per cent are neutral.

launchers, and shot-guns. Unlike other underground revolutionary movements, the Moro guerrilla units did not employ unconventional weapons, such as home-made bombs and other explosive devices. Instead, they attempted to manufacture simpler conventional weapons such as M79 grenade-launchers.

Apart from fighting men, regular villagers also played their various roles in the movement. I was told that many farmers contributed 10 per cent of their crops to the Front. Some villagers voluntarily became informers. Strangers or soldiers entering the villages could not escape their notice. The strength of the Front, therefore, cannot be measured by counting the units of guerrillas, but must also take into account the potential force of villagers. Table 3.1 shows the potential stronghold of the MNLF-Reformist Group as assessed by its Chairman.

The RSRC's activities have been mainly to consolidate supporters. Fighting is avoided unless necessary. According to Commander Iqra, in many villages, people in various occupations, such as farmers, students, and vendors, are organized into so-called 'united fronts'. Seminars and workshops are held in different *madaris* in the villages and *ustaz* and lawyers are invited to speak on various topics regarding the struggle. In some cases, Muslim foreigners are also invited to observe the Committee's general meetings and to give a talk to the villagers. Mass meetings, which usually take place in the village mosques, are often called to generate feelings of togetherness among *mujahideen* and villagers. In an attempt to become self-sufficient, the RSRC directs the *mujahideen* and the villagers to put more effort into their economic activities, to grow more crops and to raise more animals.

With respect to military activities, there had been no major clash between the RSRC units and the government forces since August 1984, when the RSRC sent its units to help a non-partisan group led by Commander Narrah fighting the Philippine Constabulary in Lumbaa Bayabao and Maguing, Lanao del Sur. One minor clash occurred in Marawi City a few weeks after I left the group, when a unit of about eight guerrillas encountered government soldiers as they attempted to collect alms from Muslims in the city. Three guerrillas and two soldiers died in the fight. During my visit, I did not see new recruits in training, but I was told that recruits are occasionally trained in several of the remote villages. Some of the RSRC leaders admit that the conflict between the various factions of the movement weakens their struggle. When asked what the main cause of the split was, Commander Iqra replied, 'Because the Moros do not have the same degree of faith in God.' The RSRC is a Maranao group of the MNLF-Reformist Group. The committee leaders, consisting mainly of secular-educated Maranao aristocrats, make their decisions independently, following the general directions of the Central Committee based in Sabah.

The MILF, Lanao del Sur, Maguindanao, and North Cotabato

With the help of the Reformist Group, I had no difficulty in making contact with the MILF leaders in Marawi City. In early January 1985, I was given the opportunity to meet a group of about 25 armed guerrillas who had come down from Camp Busrah to a nearby village in Lanao del Sur. The group was led by Ustaz Abdul Aziz, Vice-Chairman for Military Affairs in the MILF. The Ustaz and several other leaders with him were graduates of Al-Azhar University in Egypt and other institutes of higher learning in the Middle East. Most of them were *ustaz* teaching in *madaris* in the villages; some were *du-ah* employed by the centres of the Islamic call in different Arab countries. Several guerrilla commanders were still fresh from their overseas training, especially from Pakistan. With them was an officer of the PLO, apparently on a visit.

The majority of the group at Camp Busrah are Maranaos. The camp was built in an attempt to create a 'prototype community' based on the *sharia* and to uplift Islamic consciousness among the Maranaos. They believe that the strength of the Moro struggle is basically proportional to the degree of Islamic consciousness of the Moro people. They and their families in the camp attempt to follow Islamic teachings as strictly as they can. When asked about the schooling of their children, they said that they had built several *madaris* in different villages for them. 'Our children need to know two things: to understand Islam and to know how to fight,' commented one commander.

Despite their preoccupation with the preaching of Islam, the group at Camp Busrah was also fully armed and well trained. The types of weapons used were similar to those of the RSRC mentioned earlier. Camp Busrah is also a training ground for new recruits, particularly those from villages in the two Lanao provinces. Leaders at Camp Busrah retain their close contact with religious groups in Muslim countries. Several Maranaos are said to be able to secure places in universities and high schools in the Middle East through these contacts. More importantly, certain *madaris* in the villages are believed to have been built with foreign funds obtained through the efforts of these leaders. They also seem to produce disciplined commanders, some of whom are selected for further training overseas.

In mid-January 1985, I visited Camp Omar in Maguindanao and Camp Abubakar in North Cotabato. Arrangements were made by two senior religious leaders in Cotabato City for my visit to the two camps. From Cotabato City, it took many hours of driving to reach Camp Omar. We were welcomed by at least 30 armed guerrillas and by more than 50 people fetched from nearby villages. Unlike Camp Busrah, Camp Omar was not actually intended as a model community, but rather as a guerrilla training ground. It was equipped, in addition to the regular light arms such as M16 and Russian AK 47 rifles, with weapons including M1919A4 machine-guns and M2 infantry mortars. Conventional communication equipment, such as walkie-talkies, was used.

Camp Omar was commanded by a group of young military commanders of lower rank than those of Camp Busrah. Judging from the names of commanders introduced during our meeting, several are Maguindanao aristocrats. They seem to take direct orders from leaders in Camp Abubakar or in Cotabato City. Camp Omar was filled with crops such as maize and cassava; *mujahideen* and young trainees were encouraged to grow crops and to take part in economic activities which correspond to the concept of self-reliance promoted by the Front.

As in Camp Omar, a number of guerrillas and villagers gathered in Camp Abubakar on the day we were supposed to visit. Unfortunately, due to a misunderstanding, we arrived at Camp Abubakar in the afternoon of the next day and no one was expecting our arrival. We passed the outer and inner entrance checkpoints of the camp and stopped in the middle of the school playground. As we got off our 125 cc motor cycle, two men approached us, but there was no sign of other activities. The camp was quiet. There were several sheds constituting a *madrasa* (religious school) and rows of 10 to 20 huts made of bamboo. At one side of the school playground, a curtain of forest stretched to the foot of the mountains a few hundred metres away. On the other side was a stream which ran through the camp. I was told that many families lived in the camp and their children were taught in *madrasa* by two *ustaz* employed by a centre of the Islamic call in the Middle East. After about 15 minutes there were about 40 men and women gathered in a shedhouse of the *madrasa*. Half of them were guerrillas and commanders and they were fully armed. Al Haj Murad, Chief of Staff of the MILF, was unable to join us, however.

Camp Abubakar in North Cotabato is the internal headquarters of the MILF. It is the largest of the seven existing camps of the MILF. The remaining camps are Camp Omar in Maguindanao, Camp Khalid in South Cotabato, Camp Othman in Bukidnon, Camp Salman in Zamboanga del Sur, and Camp Busrah and Camp Ali in Lanao del Sur. All these camps and other 'liberated' villages operate under the general guide-lines of Camp Abubakar. As internal headquarters, it functions as a link between the Central Committee based in Pakistan or in Saudi Arabia and the various camps and provincial committees in Moroland. This, as leaders at Camp Abubakar pointed out, is the decentralized nature of the 'consultative and collective leadership system' practised by the MILF.

Camp Abubakar is dominated by members of the religious élite (*ustaz*) and religious-oriented Maguindanao aristocrats whose mission is generally to strengthen the military and economic power of the Front, to establish revolutionary committees in the provinces specified by the Tripoli Agreement, and to upgrade the Islamic and political consciousness of *mujahideen* and the Moro people. A Five-Year Plan has been initiated to build up military strength. Moro villagers from 14 to 35 years of age are recruited and trained locally and abroad. Simple types of conventional light weapons, such as the M79 grenade-launcher, are said to be manufactured. Attempts are also made to promote economic self-sufficiency. Some *mujahideen* and villagers are encouraged to engage in farming and in small-scale business activities.

While strengthening the organization, leaders and commanders at Camp Abubakar and Camp Busrah, who constitute the core members of the Central Committee, have at the same time expressed their willingness to negotiate with the Philippine government on the basis of the Tripoli Agreement. As soon as they feel strong enough to step up their military activities in order to bring about the negotiation, they, like the leaders of the RSRC, will not hesitate to do so.

The MNLF-Misuari Faction, Lanao del Sur and Sulu

In late December 1984, I made a two-day visit to a typical small Moro farming village in the foothills of the mountains in Lanao del Norte. Commander Solitario, Chairman of Ranao Norte Revolutionary Committee (RNRC), and a unit of 12 armed guerrillas welcomed us and showed us around the village. The villagers were very friendly and hospitable.

The RNRC is a Maranao group of the Misuari Faction. It consists largely of Misuari loyalists and young Maranao nationalists, particularly those who had spent some time as students or workers in Libya. Commander Solitario, a secular-educated Maranao, is himself a close friend of Misuari and had spent a period of time in Libya. Links between the RNRC and the Central Committee in Libya, therefore, have been quite strong. The group continues to receive a financial allocation from the Central Committee. Nevertheless, the RNRC makes its own local decisions based upon the general policy guide-lines of the Front.

The immediate strategy of the RNRC was to keep military activities at low key, avoiding unnecessary clashes and maintaining minimum active forces, while waiting for certain changes in the Philippine political scene. Many guerrillas returned to their normal life as farmers, students, and even government employees. The activities of the group during this 'waiting period' seemed to be concentrated on stepping up political propaganda against the Marcos regime, including publication and distribution of *Suwarah Moro* (*Voice of the Moros*). *Suwarah Moro* was not only directed against the government; it was also an attempt to disseminate the faction's political stance among the Maranaos. Nevertheless, the RNRC did not hesitate to participate in battles and sent its guerrilla units to fight in the sporadic battles between government soldiers and the joint forces of the Moro faction which took place in Lumbaa Bayabao and Maguing, Lanao del Sur, during June and August 1984 (Solitario, n.d.; *Christian Conference of Asia News*, 15 October 1984; *Malaya*, 16 August 1984).

A seven-day trip to Sulu was made in early January 1985. When I arrived, my friend, 'Wati', told me that there exist two 'governments' on the island of Sulu: the Manila government, which controlled the capital city of Jolo, and the 'MNLF government' which ruled the countryside. In other words, the MNLF-Nur Misuari Faction had solid support from the Tausug villagers in Sulu. As we drove through the outskirts of Jolo, scores of armed men in green uniform were seen walking along the unpaved road. They were identified as either guerrillas or villagers. Such scenes seemed to confirm Wati's observation. The island is divided into zones, with groups of villages commanded by different leaders. Some zones built their guerrilla camps; others did not. There are at least 13 guerrilla camps of the MNLF-Nur Misuari Faction throughout Moroland (*Far Eastern Economic Review*, 11 September 1986). Except for a few groups of bandits, villagers within these zones are under the influence of the Front and the villages are considered to be 'liberated' areas. Though one may see government soldiers on routine patrol in these villages, the soldiers do not dare interfere in the village affairs.

I visited a village on the outskirts of Jolo. It was under the command of Commander Gadil, a former aide-de-camp and a close friend of Chairman Misuari. Commander Gadil, a young secular-educated Tausug, returned from Libya in 1983 and assumed his present position as commander of one of the 'liberated' zones. He and 10 of his fully armed guerrillas welcomed us and extended hospitality. As in other 'liberated' areas, we wandered around the village without fear or hesitation. I was informed that Commander Gadil's responsibility was not so much the upgrading of villagers' political consciousness, but to consolidate different leadership groups, such as the traditional, non-traditional, and religious élites, within the movement. Attempts were also made to build a closer relationship between members of the organization and Muslim government employees. To increase the number of fire-arms in his group, Commander Gadil occasionally bought weapons from certain weapon dealers, M16 rifles cost US$600–800 and AK47 rifles, US$1,100–1,300.

Commander Gadil had several small fighting units of 8–10 men on active duty. The units are kept small for purposes of mobility and economy. Each individual unit is an autonomous entity which is free to choose its own methods of combat. Such autonomy is said to have several advantages. For example, fighters are not hampered by a set of conventional tactics, and they may 'discover' useful fighting techniques (Ruth L. P. Moore, 1981: 94–5). This unconventional approach to warfare has contributed to the Moros' reputation for being unpredictable and deadly warriors.

Like Commander Solitario, Commander Gadil continued to maintain close contact with Chairman Misuari. While consolidating the Front and keeping the villages under its influence, he was also adopting a policy of wait-and-see, waiting for the Marcos regime to be replaced before making any serious political or military commitments.

Concluding Remarks

The Moro people are warriors. They have been at war for more than 300 years and have many times fought without a formal organization or under united leadership. In the contemporary struggle, the MNLF was the first to establish an organized front with an administrative hierarchy, including a policy-making body and local operational units. In practice, however, the hierarchical structure of the organization gives way to the vagaries and flux of village life. It is one thing to organize and direct the central and provincial committees and quite another to unite and control the thrust of a mass movement consisting of different ethno-cultural groups. The MNLF is, therefore, a loosely organized front, unable to control the behaviour of the various factions or to establish a clear chain of command. Even during the height of the movement, the organization's policy-making body, the Central Committee, contented itself with furnishing broad policy outlines, giving the local leaders the power to make their own decisions. Its major function was to consolidate existing groups of Moro fighters by supplying them with weapons and funds received from various sources.

As the Front's resources decreased and fighting declined to sporadic clashes, especially after the Tripoli Agreement, local groups received less in funds and supplies than they used to. The organization became even more loosely knit and finally fissured. In other words, the Moro fighters of different groups were absorbed into the orbit of the MNLF leadership not by virtue of its institutionalized and cohesive system of organization but because of its resources and the conviction of a common cause. For example, there were no exact criteria of the Front's membership. It had no machinery of enrolment; admission was accompanied by no official formalities. Only the adherent's activity within the Front could determine the extent of participation.

As the MNLF split into three factions, the liberation movement was weakened. All factions now concentrate on consolidating power, while avoiding unnecessary encounters with the government forces. Additional provincial committees, for instance, are organized, and more camps are built. Seminars and mass meetings are held to reinforce the commitment of members and to generate group loyalties and some degree of common ideology. Economically, the factions seek to develop their self-reliance by directing *mujahideen* and villagers into economic activities.

Despite their similar strategies, the major factions are led by three different leadership groups. The MNLF-Nur Misuari Faction, which is said to be the strongest, is dominated by secular-educated non-traditional leaders. Many of the central and provincial committee members and military commanders are drawn from this leadership category. Religious leaders, such as *ustaz* and *du-ah*, rank second, while members of the traditional aristocratic élite come third. The MILF, the second strongest, is led by religious-educated traditional and non-traditional leaders. The secular-educated aristocrats are less influential. In the case of the MNLF-Reformist Group, its committee members and guerrilla commanders come mostly from the secular-educated traditional and non-traditional élites; religious leaders rank next. Unlike the Suluanos and Maguindanaos, who generally adhere to their own ethnically based factions, the Maranaos are less cohesive in support of the Maranao-based Front. They tend to join different groups on the basis of their ideological orientations and personal loyalties. In the final analysis, while the three groups are loosely knit fronts, their strength lies in the fact that they are ethnically based organizations whose leadership reflects the leadership of their respective communities.

The Malays

Before discussing the nature of the Malay separatist struggle, it is necessary to reflect briefly on the Thai government policies of integration that precipitated separatist conflict in the Patani region.

Realizing that the compulsory education policy initiated in 1921 and the cultural assimilation programme of 1939 failed to substantially promote the spread of Thai education and culture among the Malay-Muslims, the government of Field Marshal Sarit Thanarat promulgated in 1961, in connection with the overall strategy of national integration, a programme of 'educational improvement' in the four provinces. The aim was to transform the traditional Muslim religious schools (*pondok*) into registered private Islamic schools (*rongrian aekachon sonsasana Islam*). Although the converted *pondok* remained 'private', they were subjected to government regulation. The programme was designed to lay a solid foundation of Thai education and language among the Muslims in order to 'create and improve Thai consciousness, cultivate loyalty to the principal institutions such as the nation, the religion and the monarchy ...' (quoted in Pitsuwan, 1982: 194).

Under this programme, all registered *pondok* were required to teach a government-designed curriculum with the Thai language as the medium of instruction. Many secular and religious textbooks were issued by the Ministry of Education.[6] This interference by the government had a disturbing effect on the Malay-Muslim community. It destroyed the traditional *pondok* system, renowned for its religious scholarship, but the government-controlled *pondok* gradually faded away since they could no longer satisfy the Malay community with the quality of their religious education and moral cultivation. In 1984 there were 348 registered Islamic schools but only 202 were functioning (Thailand, Education Region II, 1984: 36; see Table 2.3). They are gradually being replaced by Thai public schools. As a result, more and more of the younger generation of Malays are being sent to Malaysia, Pakistan, and the Arab countries for their Islamic as well as secular studies. Many Patani Muslim students overseas receive financial assistance from foreign sources and develop an increased sense of antipathy towards the Thai state. These foreign students are potential leaders of the various separatist fronts.

In addition to its educational programme, the Sarit government initiated a project called *nikhom sangton-eng* to redress the population imbalance in the Muslim areas, as discussed in Chapter 1. The determination of the Thai government to carry out these assimilation programmes, together with the 'internal colonial' characteristics of the existing socio-economic and political structure of the Muslim society described in the previous chapters, explain the rise of separatist politics and the emergence of different liberation fronts.

The Malay Liberation Fronts: BNPP, BRN, PULO, and BBMP

There are four major Muslim underground groups currently active in the four southern border provinces of Thailand. They include the Barisan Nasional Pembebasan Patani (BNPP) or National Liberation Front of Patani; Barisan Revolusi Nasional (BRN) or National Revolutionary Front; Patani United Liberation Organization (PULO) or Pertubuhan Perpaduan Pembebasan Patani; and Barisan Bersatu Mujahideen Patani (BBMP) or United Fronts of Patani Fighters. As discussed in Chapter 1, Malay resistance transformed from a movement largely organized around members of the Malay ruling élites to one with a broad base of popular support. This transformation resulted in the formation of GAMPAR in Malaya and the PPM in the Patani region. The two organizations co-ordinated their operations closely. Their struggle, however, was confined to political activities, with only sporadic outbursts of violence, such as the uprisings in Kampung Belukar Samok in 1947 and in Dusun Nyor in the following year. When the GAMPAR and PPM leaders died in 1953 and 1954 respectively, the organizations disintegrated. Drawing former members of the defunct GAMPAR and PPM, Tengku Abdul Jalal (Adul Na Saiburi), former deputy leader of GAMPAR, formed the BNPP in 1959 (BNPP, 1981a: 5).

The BNPP leadership comprised both traditional aristocrats and religious leaders. Its pattern of resistance became more ideologically inclined; its objective was no longer autonomy or irredentism but restoration of independence. The preferred strategy also was widened to include not only political activities but also armed guerrilla warfare. Idris bin Mat Diah, alias Pak Yeh, a renowned gang leader in the Muslim provinces, joined the Front and became a guerrilla leader. Surrounded by Muslim gang leaders, former bandits, and outlaws, Pak Yeh began his guerrilla operations in the four provinces. From the early

1960s, armed clashes with the government forces occurred intermittently.

The early 1960s also saw Muslim religious students, particularly those who studied in Malaya, join the BNPP. Some of these students, such as Badri Hamdan, Abdul Fatah Omar, Hannan Ubaidah (present leaders of the BNPP), continued their studies in Al-Azhar University in Cairo. While in Cairo, they formed an organization known as Rumah Patani (House of Patani) which became the BNPP's overseas base. In Mecca, an association called Akhon (Brother) was organized by Patani students and workers, with the aim of promoting education of Patani Muslim youths. However, Akhon soon became a training ground for young Patani activists who supported the liberation politics back home. Many of the present leaders of the BNPP in Saudi Arabia, Egypt, and Pakistan are former leaders of the dissolved Akhon.

At the same time, rifts among the Muslim activists emerged. The BNPP's ideological commitment at this early stage was not as clear as it became in the 1970s. Although its objective was to gain independence, the idea of restoring a sultanship was still floated by some of its leaders and supporters, especially the former aristocrats. This left the impression that should the struggle of the BNPP succeed, the Patani sultanate might be restored. The more progressive Muslims, such as Ustaz Abdul Karim Hassan and his colleagues, admirers of Sukarno of Indonesia, hesitated to join the BNPP. In March 1963 Ustaz Abdul Karim Hassan and his group formed the BRN with the aim of establishing a Republic of Patani (interview with Ustaz Abdul Karim Hassan, Kuala Lumpur, Malaysia, 12 June 1985). The main factor contributing to the emergence of the BRN was ideological differences.

The BRN leaders placed greater emphasis on political organization than on guerrilla activities. Their strategy was to penetrate the *pondok*. Because the Chairman himself was the headmaster of an Islamic school, the task of penetration was not very difficult for the BRN. Within five years (1963–8), the BRN was able to influence several of the Islamic schools, especially in Narathiwat and Yala provinces. In 1968 Ustaz Abdul Karim Hassan and other BRN leaders went underground. There had been little military activity throughout the 1960s except for sporadic encounters with the BNPP's guerrilla units, which at this stage were treated by the Thai authorities as bandit gangs rather than as disciplined, armed insurgent units.

Also in 1968 the third front, PULO, emerged (PULO, n.d.: 1). The PULO was first organized in India by Tengku Bira Kotanila (Kabir Abdul Rahman), who had at the time just completed his first degree in Political Science at Aligarh Muslim University, and a group of Patani students at the same university. Soon after the formation of PULO, Tengku Bira Kotanila moved to Mecca and focused his recruitment efforts primarily on young, non-committed Patani Muslims by stressing secular nationalism, and differentiating PULO from BNPP's orthodox Islam and BRN's 'Islamic socialism'. The PULO also cultivated its support in the homeland, in Malaysia, and among Patani students in various Arab countries.

When Tengku Abdul Jalal, Chairman of the BNPP, died in September 1977, the Central Committee elected Badri Hamdan to lead the Front.[7] Badri Hamdan has brought several changes to the organization. He has, for example, attracted many religious graduates of the Middle East universities to form a more aggressive, Islamic leadership cadre and has penetrated Islamic schools to broaden the base of popular support. In 1985 eighteen Islamic schools were claimed to have been administered by the BNPP (information obtained from the BNPP headquarters, Malaysia, 19 February 1985). Under the present leadership, more systematic political and military training is organized. Moreover, religious-educated leaders dominate the BNPP; the non-religious leaders, such as secular-educated and former aristocratic élites, seem to have gradually moved away from the Front; this is especially true of those who have leadership ambitions. In 1985 Wahyuddin Muhammad, former Vice-Chairman of the BNPP, and several other leaders formed the BBMP.

Although the fronts differ in many ways, they all view the Thai administration as a colonial power with which no compromise is possible and emphasize independence through armed struggle (Haemindra, 1977: 86). Since the early 1970s, the fronts have conducted guerrilla activities. Each organization developed its sphere of influence in different areas throughout the provinces. Muslim villagers who lived within these spheres were directly or indirectly involved in the activities of the fronts. They were often reminded of their obligation to be involved in the struggle and were cited the Hadith that says, 'The best of the believers is he who fights in the cause of Allah with his wealth and his life' (narrated by Bukhari and Muslim). Even though not all Muslim villagers were influenced by the fronts, most of them were aware of the secessionist politics.

At the height of their guerrilla activities (1970–5), the term 'invisible governments' was used by some Muslim villagers to describe the fronts; the provinces of Patani, Yala, and Narathiwat were sometimes referred to as the 'free region' (Haemindra, 1977: 88). The BNPP, BRN, and PULO set up ambushes and attacked police check-points and government installations. Extortion, especially from rubber and coconut plantation owners, and kidnapping for ransom were also part of their activities. Local Thai businessmen had to pay 'protection money' in order to live in peace. Terrorism was often used to remind the villagers and the government of the fronts' existence (for an account of violent activities in the Muslim provinces during 1976–81, see Satha-Anand, 1987).

Between 1968 and 1975 the Bangkok government launched a series of military operations against Muslim 'terrorist' activities. The operations were code-named 'Special Operation for the Four Southern Border Provinces', 'Ramkamhaeng Operation', and 'Special Terrorist Campaign in the Three Border Provinces' (Megarat, 1977: 21–2). They involved Police Special Forces of the Ninth Zone and required large-scale recruitment of Buddhist and Muslim volunteers. The operations were also incorporated into regular army and police force operations in the area. The result of this seven-year campaign was as follows: 385 clashes with Muslim 'terrorists'; 329 terrorists dead; 165 surrendered to Thai authorities; 1,208 arrested; 1,451 weapons of various types, 27,538 rounds of ammunition, and 95 grenades captured by the authorities; and 250 terrorist camps destroyed (Megarat, 1977). The then Police Captain Megarat, one of the participants in the operations, made the following comments: 'If we look at the statistics, we like to believe that our operations were successful, and the terrorists should have been entirely wiped out. On the contrary, several terrorists remain active; new leaders who are unfamiliar to us have appeared. In fact, we have conducted campaigns against them since 1905. Yet, they can still exist' (Megarat, 1977).

Indeed, the Muslim separatist movement continues to operate, and the conflict appears to have deepened and escalated. The murder of five Muslim youths, allegedly by Thai soldiers, in December 1975 sparked the largest demonstration in the history of Patani. Thousands of Muslims gathered at the Patani Central Mosque each day for 45 consecutive days. During this period at least 25 other Muslims were killed and some 40 injured, again allegedly by Thai armed forces (Suthasasna, 1976: 127–8). This demonstration, however, can also be viewed as being related to the general political awareness during a three-year interlude of democracy (1973–6) in Thailand.

Some separatist leaders interpreted this demonstration as an expression of Muslim anger towards Thai-Buddhists. They believed that this anger could be manipulated and directed 'correctly' to support the struggle. More importantly, perhaps, they assumed that popular support was always potentially present and that this event was an example of the people's support for the issues the fronts struggled for. Thus, they decided not to devote their relatively weak resources to domestic mobilization at the present time. In fact, since the mid-1970s, all three fronts have placed greater emphasis on obtaining external recognition and support. Because of this external emphasis, there has been a lull in activities in the region. Nevertheless, the fronts continue to engage in guerrilla operations to make their presence known (Che Man, 1985: 108). According to a Thai parliamentary commission's report, there were 161 armed clashes between the so-called terrorists and bandits and Thai authorities in the three provinces of Patani, Yala, and Narathiwat between November 1978 and October 1979.

Externally, the fronts successfully mobilized Patani students and workers in foreign lands, particularly in Malaysia, Saudi Arabia, Egypt, and Pakistan. These Patani Malays abroad played a significant role in establishing contacts with the authorities and political parties of other countries. They were anxious to be recognized by Muslim countries as fighting in the cause of Islam and preserving the Muslim *ummah*, and not as bandits nor communists.

The initial step toward Arab recognition was made when Arab newspapers and magazines in several countries in the Middle East began to publish articles and stories about the struggle of the Patani Muslims in southern Thailand. In July and August 1976, for example, one of the largest Libyan daily newspapers, *Al-Jihad*, published a series of articles for 43 consecutive days describing the activities of the guerrillas in Patani and the problems facing them. Several books in Arabic on the Patani movement appeared, including *Patani: A Difficult and Concealed Revolutionary Movement* by Muhammad Warieth and *Islamic Government in Patani* by Dr Rauf Saliby. Journalists and Muslim officials started to visit the Patani region to obtain firsthand information. This publicity and personal observations contributed to greater Arab awareness and sympathy for the Malay-Muslim cause.

Organization

Like those of the Moros, the Malay liberation fronts are loosely structured organizations, even though complex organization charts are constructed on paper. The constitutions (*perlembagaan*) of the BNPP, PULO, and BBMP, for instance, devote many sections to the structure of their respective organizations (BNPP, 1981b: 12–21; PULO, n.d.: 3–5; BBMP, n.d.: 3–7). The BRN exhibits a similar complex organizational structure, though it publishes no documents on the matter. (Appendix 8 illustrates the organizational structures of the fronts.)

The basic components of the fronts' organizational structure are the Central Committee (Majlis Tertinggi), Provincial Committee (Majlis Wilayah) and Working Committee of the foreign branch (Majlis Cawangan). The Central Committee, which is headed by a Chairman or a Secretary-General (PULO), is the heart of the organization. Important policies and appointments are made by this body. It comprises about 13–25 members and different sections which perform various functions (BNPP, 1981b: 12–13; PULO, n.d.: 3; BBMP, n.d.: 3–4). All members of the Central Committee are appointed by the Chairman or Secretary-General. Except in the case of the BBMP, which has neither Chairman nor Secretary-General, the members of the Central Committee are appointed by the Congress (Majlis Mujahideen). The head of each section and his deputy are ex-officio members of the Central Committee. Table 3.2 shows the current members of the BNPP Central Committee. Members of the BNPP Central Committee are dominated by religious leaders, most of whom were educated in the Middle East. This pattern occurs also in the BRN and PULO. The BBMP leaders, however, are drawn primarily from secular and religious teachers who were educated in Malaysia and Indonesia.

Despite the fact that the central committee of each front operates inside the Malaysian border, it still plays a major role in carrying out local activities. Unlike those of the Moros, the provincial committees take direct orders from the central committee. Thus, the activities of the armed guerrilla wing of each front are directly under the supervision of the military section of the central committee which is responsible for the overall guerrilla operations and recruitment. Recruitment of guerrillas is

TABLE 3.2
Members of the Central Committee of the BNPP, as of 3 August 1986

Name[1]	Portfolio	Classification
Badri Hamdan	Chairman and Head of Military Section	Religious teacher
Cikgu Noor	Deputy Head of Military Section	Secular teacher
Saifullah Siddik	Vice-Chairman and Head of Islamic Call Section	Religious teacher
Abu Ubaidah	Secretary	Secular teacher
Muhammad Najib	Treasurer	Religious teacher
Shahrir Abdullah	Head of Economic Section	Religious teacher
Ibn Al-Walid Al-Khalid	Head of Political Section	Religious teacher
Hannan Ubaidah	Head of Interior Section	Religious teacher
Salahuddin Tarmizi	Head of Foreign Section	Lawyer
Zulkifli Abdullah	Head of Information Section	Religious teacher
Arifin Abdullah	Head of Education Section	Religious teacher
Jamaluddin Ismail	Head of Youth and Welfare	Religious teacher
Saifuddin Khalid	Special Duty to the Office of the Chairman	Religious teacher

Sources: BNPP (1981b); Personal communication with Badri Hamdan, Chairman of the BNPP, 6 August 1986.
[1] All names mentioned in this table are pseudonyms used by the individuals concerned.

normally initiated by guerrilla leaders who propose individual candidates to the military section. The names of the selected candidates are then submitted to the Chairman for approval. For example, Article 20, Section 8 of the BNPP's constitution states, 'Permission must be obtained from the military section to recruit armed fighters' (BNPP, 1981b: 35). Recruitment of guerrillas is also made by different section heads and members of the provincial committees; they recommend the qualified candidates to the military section for further action.

Ordinary members and supporters of the fronts are usually recruited from various categories of the Muslim population by members of the fronts. Caution is taken by all fronts to avoid infiltration by unwanted persons such as government agents, drug addicts, and individuals having ideologies inconsistent with Islam. In the case of the BNPP and BRN, recruiters are primarily religious teachers who control the Islamic schools in the provinces. They recruit teachers, students, and Muslim villagers in the vicinity of their schools.

The fronts have yet to develop specialized training programmes for different functions such as propaganda and intelligence. Generally, the fronts perform two training programmes: military and political. Guerrillas are trained locally and abroad. Local training is conducted by local guerrilla leaders in remote areas in the foothills of the mountains. Local training is, however, not scheduled regularly; it depends mainly on the budget and the demand. Arrangements are also made with Muslim countries such as Libya and Syria for military training. A group of young guerrillas from the various fronts is from time to time sent abroad for the purpose. In 1985, for instance, a group of BNPP fighters was given several months' military training in Afghanistan (interviews with several guerrillas returning from Afghanistan, Malaysia, 6 May 1986).

Political training is conducted by religious teachers who are normally high-ranking leaders of the fronts. In the case of the BNPP, political training is in the form of orientation. Selected groups of 10–15 recruits are provided with food and lodging for two to three weeks. During this period, lectures on different topics are given by the BNPP leaders and invited religious teachers. The programmes are aimed at familiarizing the recruits with the objectives, ideology, and discipline of the Front. The recruits return to their respective villages and are asked to perform certain duties such as recruiting, collecting information, and serving as couriers. The programmes started in 1979. (The author observed training programmes in May 1981 and in June 1985.)

Financial resources come mainly from the contributions of local and foreign members and supporters. Contributions are regularly made by members who hold permanent jobs. In the case of the BNPP, each full-time employee contributes according to his capacity, from US$4 to US$80 monthly. Ordinary members are required by the constitution to pay a due of US$0.40 monthly (BNPP, 1981b: 21; 1982). In addition, members are encouraged to provide food and other necessities to guerrillas when the units move to their vicinities.

A second source of funds is from voluntary and non-voluntary local supporters. Voluntary supporters are Malay-Muslims of various occupations who support the fronts because they believe in the struggle or because they want to protect their own interests. The non-voluntary contributors, who are mostly non-Muslims, include local businessmen, plantation owners, and shopkeepers.

A third source of income is derived from members abroad, especially in Saudi Arabia and Malaysia. The BNPP, PULO, and BRN receive regular contributions collected from about 15,000 Patani workers in various cities in Saudi Arabia. Patani workers in other countries contribute similarly. Many Patani Muslims who reside in Malaysia contribute money and knowledge to their preferred fronts.

But the fronts' major source of finance is contributions made by various agencies, organizations, and wealthy families in Muslim countries. Most funds are given in the form of charity. The BNPP, for example, has obtained a *fatwa* from Sheikh Abdul Aziz Ibn Baz, President of the Department of Scholarly Research and Religious Ruling of Saudi Arabia, saying that the Front has met the requirements for acquiring charity. The BNPP, therefore, has been collecting alms in Saudi Arabia. It is believed that the BNPP has also received some donations from the Islamic Solidarity Fund through the Islamic Development Bank (IDB). The authorities of the IDB deny it, but admit that the IDB has allocated several scholarships for Muslims in Thailand (interviews with IDB authorities, Jeddah, Saudi Arabia, April 1986). The Muslim World League, too, often makes donations to the BNPP. But the primary source of income of the BNPP is said to be the Al-Auqaf (Welfare Department) and Islamic Call Society in Kuwait.

The fronts spend substantial amounts of their income on operations and management. A fairly large portion of their finance is used for welfare purposes such as building schools for the children of *mujahideen* (BNPP, 1981c, 1981d).

The fronts share certain common objectives: to fight for the independence of Patani; to establish a state and a society based on Islamic values and way of life; to promote Islam and preserve Malay identity and culture; and to respect and abide by the UN Charter (BNPP, 1981b; PULO, n.d.; BBMP, n.d.). All fronts stress the Islamic obligation of *jihad* and Malay nationalism. Despite sharing common objectives, the fronts differ in ideology. The BNPP[8] is a conservative group committed to orthodox Islam. As stated in its constitution, 'the basic political ideology of the BNPP is based on Al-Quran, Al-Hadith, and other sources of Islamic law' (BNPP, 1981b: 4–5). In contrast, the BRN is a more radical group and is most outspoken in its opposition to the feudal institution of the sultanship. It has often been stated that the BRN ideology is 'Islamic socialism'. In the early 1970s the BRN was believed to have co-operated with Rashid Maidin's faction of the Communist Party of Malaya (CPM) (*Asiaweek*, 4 April 1980). However, neither Rashid Maidin's faction nor the BRN has ever made any public comments on the matter. When Ustaz Abdul Karim Hassan, then Chairman of the BRN, was interviewed in June 1985, he denied any co-operation between his organization and the CPM. Because there has been no public statement by the Front on ideological matters, it is difficult to explain what is meant by 'Islamic socialism'. Some Patani Muslims argue that there is no such thing as 'Islamic socialism' or 'Islamic capitalism'. This is because Islam, as a system in itself, can only retain its original property when it does not blend with non-Islamic systems; if it does, it can no longer be regarded as Islam. Nevertheless, a BRN spokesman explains that 'Islamic socialism is more or less similar to what was practised in Egypt under Gamal Abdel Nasser' (interviews with several BRN leaders, Patani, February 1980).

The constitution of PULO states, 'Ideology of this organization is based on *IBANGTAPEKEMA* which means Islamism, statehood, nationalism, and humanitarianism' (PULO, n.d.: 2). The constitution further explains that Islam, democracy, equality of opportunity, and humanitarianism constitute the socio-political and economic principles of the Front (PULO, n.d.). Ideologically, the PULO stands in the middle between the conservative BNPP and the radical BRN. However, it often sways either way to suit its purpose. The basic ideology of the BBMP is also Islamic (BBMP, n.d.: 1). Since it was founded only in 1985, it is too early to discuss the nature of the organization or to predict its future performance. As a new organization, the BBMP concentrates on propaganda activities, distributing documents and pamphlets to gain acknowledgement of its existence. Calls for meetings with leaders of other fronts have also been initiated by the BBMP in an attempt to find common ground for co-operation (BBMP, 1986).

In addition to ideological differences, members and supporters of the fronts also come from different groups of people. The BNPP leaders and supporters are generally conservative religious teachers, intellectuals, and members of well-off families. It is common knowledge among the Patani liberationists that the BNPP is an organization of the 'upper class'. Many of its members are of the opinion that socialism and liberal democracy are inconsistent with Islamic teachings. The BNPP revolves around conservative Muslim countries like Saudi Arabia, Kuwait, Malaysia, and Pakistan. The BRN draws most of its leaders and supporters from the liberal religious élite and nationalists who believe that Islamic socialism is the most appropriate form of government. The BRN seeks close ties with liberal Muslim countries such as Algeria, Syria, and Libya, but concentrates its activities in Malaysia and Saudi Arabia. On the other hand, leaders and supporters of the PULO come primarily from young militant nationalists and former gang leaders. The Front is more at ease with militant Muslim states such as Syria, Iran, Algeria, and Libya. Its guerrillas are from time to time sent to be trained by the PLO (*Arabia*, August 1982). The PULO's attraction to the PLO is, however, less a result of common ideological inclination than the PLO's militancy.

Although the Malays are not divided along ethno-cultural lines like the Moros, factional conflicts within the Malay struggle occur on the basis of ideological and class differences, as discussed above. Several attempts have been made to unite or at least to work together in facing the common enemy. The most significant attempt at unity was initiated by the Muslim World League in 1979, when the BNPP and PULO signed a short-lived agreement to co-ordinate their military operations. During this period of co-ordination, the fronts were reportedly fielding considerable numbers of armed guerrillas. In late 1979 a Thai parliamentary commission on southern Thailand identified 84 different armed groups in the four provinces of Patani, Yala, Narathiwat, and Songkhla (*Far Eastern Economic Review*, 20–26 June 1980). After a year of co-ordination, the two fronts could no longer find any common ground to enable them to compromise their differences. They terminated their agreement. A recent attempt at unity was made in June 1986, when the BBMP invited the other three fronts to a meeting in Kuala Lumpur to discuss possible common ground for co-operation and unity among them (BBMP, 1986: 1–2). The BNPP refused to accept the invitation, accusing the newly created BBMP and the recently installed BRN and PULO leaders of seeking to gain recognition through the unity forum.

In recent years, the fronts have also been plagued by internal rifts. In 1980 the top two leaders of the BRN, Ustaz Abdul Karim Hassan and 'Hajji M.', were unable to compromise their differences on several issues. Their disagreement divided the BRN into two factions. In 1984 Ustaz Abdul Karim Hassan was forced to hold an election in which he lost his leadership to younger-generation leaders, Cikgu Peng (Pak Tua) and Pak Yusof. Cikgu Peng was elected Chairman while Pak Yusof was appointed Secretary. However, Ustaz Abdul Karim Hassan still considers himself to be the leader of the BRN; he and his followers have dissociated themselves from the newly installed leaders (interview with Ustaz Abdul Karim Hassan, Kuala Lumpur, 12 June 1985). This means that the Front has split into three groups.

In early 1981 a rift occurred within the PULO between a group that favoured co-operation with bandit groups and those who opposed it. The dispute resulted in the expulsion of many leaders of the latter from the organization and divided the PULO into two factions. Also in 1981 a rift within the BNPP emerged between a group of religious teachers loyal to Badri Hamdan and those secular leaders who revolved around Wahyuddin Muhammad, Badri Hamdan's deputy. The split reached its peak when the latter formed the BBMP in 1985 (BBMP, n.d.). Thus, the BNPP, which had hitherto managed to keep internal division under control, fissured.

Apart from ideological differences, the fronts also differ in tactics and in strength. From the late 1960s to the mid-1970s, the BNPP organized a comparatively strong guerrilla group called the National Liberation Army of the Patani People (NLAPP). It consisted of several hundred men recruited from different areas of the provinces. The BNPP became rather well known as a result of the NLAPP's militant activities, though the Front was at the same time weakened by the loss of members and supporters in battles, arrests, and surrenders. Consequently, the BNPP leaders of the 1980s feel that they have not only to recoup the Front's strength, but also to gather more support before launching another phase of militancy. The current emphasis is, therefore, on political organization and developing external political contacts rather than conducting insurgent activities. Whether or not this stance has led to a further weakening of the Front is debatable. Nevertheless, the BNPP still claims to have several small units of seven to fifteen guerrillas stationed in various districts in the provinces of Patani, Yala, and Narathiwat. They are armed with conventional weapons including M1 carbines, M14, M16 and AK 47 rifles, M1919 machine-guns, M79 grenade-launchers, shot-guns, pistols, and home-made bombs. The guerrillas were commanded by Pak Yeh until he died in 1984. Presently, all guerrilla units are under the centralized command of the military section headed by Badri Hamdan, Chairman of the Central Committee. Under Badri Hamdan, guerrillas are trained more regularly and better fighters are produced. This is because the BNPP in recent years has received greater financial assistance from abroad. The BNPP, moreover, has been able to secure overseas grants for certain Islamic schools and mosques in the provinces.

Under Ustaz Abdul Karim Hassan's leadership, the BRN placed emphasis on widening the base of internal support while organizing some guerrilla units. Although the BRN had fewer fighting units than either the BNPP or PULO, a well-equipped guerrilla group of the BRN was active in the mid-1970s. The group disintegrated when its courageous leader, Cikgu Din Adam, was killed in 1977. The Front also formed several small but active units of urban guerrilla forces. However, the BRN under the leadership of Ustaz Abdul Karim Hassan was better known for its ability to mobilize support through the Islamic schools.

Of the three fronts, the PULO was most militant. It shares the political idea that 'Political power comes out of the barrel of a gun' (Mao and Guevara, 1962: 13). When the BNPP leaders began to emphasize organizational work, the PULO accused the BNPP of being weak and stalling on liberation. The PULO did not hesitate to co-operate with bandit groups so as to create as much trouble for the Thai authorities as possible. It is not surprising that PULO was constantly in the news, being accorded responsibility for kidnapping, ambushes, and bombing.

The PULO used its office in Mecca as a 'policy-making headquarters' and that in Malaysia as an 'operations office'. Important decisions were conveyed from Mecca to Kota Bharu, about 30 km from the Thai border, where instructions were given to leaders and guerrillas. Another major office of PULO was in Damascus where selected guerrilla leaders were often sent for military training. In 1981 the PULO claimed to have about 20,000 members who contributed US$0.40 monthly. It was also acknowledged to have many units of guerrillas stationed throughout the three provinces. However, the PULO faced its worst crisis in 1984. Its headquarters in Mecca were raided by Saudi authorities. Many of the Front's leaders were arrested or deported. More than 700 of its members were sent home (interviews with several informed Patani Muslims, Mecca, Saudi Arabia, April 1986; Centre for Administration of the Southern Border Provinces, n.d.). Tengku Bira Kotanila, Secretary-General of PULO, escaped arrest but was forced to resign. 'Ustaz Abdul Hadi', Tengku Bira Kotanila's deputy, assumed the leadership. The reason behind the Saudi's harsh action against the PULO was that the organization's activities had gone beyond the limit that the Saudi government could tolerate. For example, the PULO openly issued a 'Citizenship Identification Card of Patani Republic' to Patani Muslim workers in Saudi Arabia and collected taxes from them. Furthermore, its constant contact with the officials of the Baath Party of Syria and the opening of its new office in Iran irritated the Saudi government. The crisis has undoubtedly crippled the PULO, and it may take several years before it can revive its activities in Saudi Arabia. The difficult task of restoring the organization to its original strength is now in the hands of younger-generation leaders headed by 'Ustaz Abdul Hadi'.

Finally, internal cleavages have weakened each front in recent years. Of the three fronts, the BNPP seems to be least affected by such cleavages, though some of its leaders have joined the BBMP. The Front's political and military training programmes, initiated in 1979, have continued with little interruption. More importantly, the BNPP has recently received increased financial assistance from Saudi Arabia and Kuwait. In contrast, the BRN has split into three groups. The present leadership of Cikgu Peng and Pak Yusof faces a difficult task in reconciling these factions. While they still advocate 'Islamic socialism', Ustaz Abdul Karim Hassan's group now advocates pure Islam (interview with Ustaz Abdul Karim Hassan, Kuala Lumpur, 12 June 1985). Reconciliation with the Hajji M. Faction is also difficult because Cikgu Peng and Pak Yusof incline toward military action whereas the former emphasizes political organization. In the case of PULO, the organization has not yet recovered from its crisis in 1984. It remains to be seen whether 'Ustaz Abdul Hadi' and his group are capable of repairing the damage that has put the Front on the verge of disintegration. At the moment, therefore, all fronts seem to be concentrating their attention on repairing the damage caused by internal cleavages rather than on keeping their liberation activities going. Nevertheless, a spate of violence since the early months of 1988, which occurred after a relative lull in the past several years, is attributed to an attempt by the fronts to remind the Muslim villagers and the Thai government of their existence.

Field Studies

In 1985–6 the author visited the offices of the BNPP in Malaysia and Saudi Arabia. The visits were confined to the BNPP because it was the only front accessible at the time. Moreover, as argued above, the BNPP is more powerful and less fragmented than the other fronts. The following account gives some information on these visits.

In mid-June 1985, I made a seven-day visit to the BNPP headquarters located within the Malaysian border. The visit was possible because some of the BNPP leaders, including the Chairman, Badri Hamdan, were my school-mates during our years in Saudi Arabia and Egypt. For the headquarters of an underground front, I was surprised to see that the office was furnished with relatively expensive furniture and office facilities such as telephones, typewriters, and copying machines. Several of the Front's leaders, such as the Chairman, Vice-Chairman, Secretary, and heads of the political and interior sections, come to the office regularly. When the Chairman is at his desk, he is usually busy with paperwork and long-distance telephone calls. The Vice-Chairman spends much of his time writing pamphlets and propaganda, both in Malay and Arabic. The Head of the Political Section concentrates on analysing local and international political developments. He is regarded as the organization's theoretician. The Secretary and Head of the Interior Section, on the other hand, are responsible for the management of guests and visitors who come and go regularly. Most of them are local leaders who visit the headquarters for various purposes. Despite the fact that communication between the headquarters and local leaders is relatively well established, leaders from the headquarters occasionally cross the border to supervise the local leaders.

Apart from office work, the Chairman and other leaders are engaged in carrying out political orientation and military training. I was permitted to observe a lecture given by the Chairman to a group of about 15 selected recruits who were attending a two-week orientation programme. The lecture essentially revolved around a Quranic verse (VIII: 60) which states:

Against them make ready your strength to the utmost of your power, including steeds of war, to strike terror into (the hearts of) the enemies, of God and your enemies, and others besides, whom ye may not know, but whom God doth know. Whatever ye shall spend in the Cause of God, shall be repaid unto you, and ye shall not be treated unjustly.

Such lectures are given on different topics by other leaders and invited 'lecturers' until the programme is completed.

During the week of my visit, the Chairman and several leaders in the Military Section were absent from the office for two days. I was told that they had gone to a camp where a group of 12 recruits undertook a two-week guerrilla training programme. The programme was initiated in 1984 in an attempt to familiarize selected members with conventional and home-made weapons. The political and military trainees will be regarded as active members of the BNPP when they return to their respective villages. However, they are seldom promoted to high-echelon positions because the BNPP prefers its policy-making leaders to be drawn from university graduates. To some extent, this has moved the Front away from the grass-root supporters.

Besides dedication to their office work, leaders and members of the Front at its headquarters are also devoted to pious duties (*ibadat*). They spend considerable time each day performing religious rituals. 'Only by being close to God', said one leader, 'can the objectives of our struggle be fulfilled.' To be close to God, they further assemble at least once a week to discuss different aspects of Islam. By doing so, they hope to increase their Islamic knowledge and, more importantly, to gain enlightenment from God. At the end of week-long activities, a meeting of members of the Central Committee is called. Decisions made at this meeting are final. Orders are conveyed to the concerned parties in the provinces and abroad for implementation.

No fire-arms are allowed at the headquarters of the BNPP. Guerrillas visiting the office must leave their weapons inside the Thai border. This is to comply with the Malaysian laws which prohibit its citizens carrying and possessing fire-arms. To the outsiders, the headquarters of the BNPP appears more like a business office than an underground organization aimed at separating the southern border provinces of Thailand.

In April 1986, I made a visit to Saudi Arabia for a period of a month. The trip was an attempt to look at the activities of different fronts in the kingdom. Saudi Arabia has always been regarded as a centre of religious education and employment for the Patani Muslims. During the late 1960s and early 1970s, when the immigration laws of Saudi Arabia were still relaxed, there were as many as 30,000 Patani Muslims living legally and illegally in various cities of that country. At least a third of that number still remains, and most of them are involved in the activities of the fronts (interviews with several informed Patani Muslims in Mecca and Jeddah, Saudi Arabia, April 1986). My trip was sponsored mainly

TABLE 3.3
Members of the Working Committee of the BNPP in Mecca
Who Were Holding Portfolios in 1986

Name[1]	Portfolio	Classification
Amin Hamdi	Chairman	Religious teacher
Abu Iman	Secretary	Religious teacher
Khalid Abdullah	Registrar	Businessman
Fathulmubeen Omar	Treasurer	Businessman
Khairullah Bakri	Officer in Charge of Office	Religious teacher
Jamal Abdul Nassir	Education Officer	Religious teacher
Hamdi Khalid	Information Officer	Religious teacher

Source: Information from the office of the BNPP in Mecca, Saudi Arabia, April 1986.
[1]All names mentioned in this table are pseudonyms used by the individuals concerned.

by the Muslim World League in Mecca. While in Mecca, however, I was provided with lodging and transportation by the BNPP leaders. A trip to Jeddah, Medina, and Taif also was arranged by them.

In 1984 the BNPP built a permanent office in Mecca, at a cost of US$70,000. The BNPP branch in Saudi Arabia is governed by a 17-member Working Committee (Majlis Kerja), seven of whom are holding portfolios as shown in Table 3.3. The Front draws its members from Patani Muslim intellectuals, businessmen, and workers living mainly in the cities of Mecca, Medina, Jeddah, and Taif. They make financial and other contributions to the organization according to their capability. Through their generous contributions, the BNPP in Mecca has persisted and even grown. It has also provided financial and material support to the Front's headquarters (interviews with Amin Hamdi, Chairman of the Working Committee of the BNPP, Mecca, April 1986).

Like the headquarters in Malaysia, the BNPP office in Mecca is well equipped. Amin Hamdi, Chairman of the Working Committee, and Khairullah Bakri, Officer in Charge of the Office, come to the office daily. The primary job of the former, apart from administrative matters, is to collect monthly dues from members and to raise alms from Arab sympathizers. The responsibility of the latter, in addition to office maintenance, is to arrange official meetings, circulate information among the members, and to lead prayers. A meeting of the Working Committee takes place once a week to evaluate overall performance and to discuss the latest developments in the homeland. Moreover, the BNPP holds two classes every week on different religious subjects, ranging from Islamic jurisprudence to Quranic interpretation. They are given by senior religious teachers. Participants in these classes are mostly leaders and members of the Front. They participate not only to widen their horizon of Islamic knowledge but also to attain enlightenment from God. Some senior religious teachers have, since 1984, been meeting at least once a week in an effort to complete a dictionary of Patani-Malay language. Another function of these senior religious teachers is to provide *fatwa* for the BNPP when required.

During my stay in Mecca, the BNPP called special meetings of different groups of its members to discuss the current situation in the Patani region. It also invited Salamat Hashim, Chairman of the MILF, to brief them on the Moro struggle after Marcos. At one of these special meetings, the Working Committee members made a decision to buy a piece of land in Patani valued about US$65,000, to be donated to the BNPP. The land is intended for growing crops to provide additional income for the Front in the future.

The activities of the BNPP in Mecca are intensified when its Chairman or Vice-Chairman from the headquarters makes his yearly visit to the Mecca office. During his visit, the Chairman will travel to major cities of Saudi Arabia to seek support from various government agencies and groups of Arab sympathizers. He will also visit several other Arab countries, such as Kuwait and Egypt. Before returning to the headquarters, a general meeting at the Mecca office will normally be called to discuss the outcome of his tour and to hear opinions and suggestions not only from leaders and members of the Front in Saudi Arabia but also from the invited representatives of other BNPP foreign branches (Egypt, Kuwait, Sudan, and Pakistan).

I also interviewed some leaders of the BRN and several former PULO leaders. The activities of the BRN in Saudi Arabia are similar to those of the BNPP. Both are active fund-raisers. Because of their financial contributions to their respective Fronts, they also play an important role in shaping the policies of the organizations. However, the two differ in that the BNPP seems to have gained more support from Patani Muslims and Arab sympathizers in the kingdom. This is due mainly to the fact that conservative Arabs and religious-educated Patani Muslims are less attracted to the socialistic inclination of the BRN ideology. PULO is no longer active in Saudi Arabia, after the events of 1984.

Concluding Remarks

Long-standing resistance and resentment against Thai assimilationist policies contributed to the emergence of the Malay liberation movement, which in turn gave rise to organized fronts with a broad base of popular support. Along with this transformation, a transfer of leadership from former aristocrats to the religious élite occurred as the prestige of the former in Patani society declined. The involvement of the Muslim religious élite in the liberation struggle has been crucial. The religious élite, especially religious teachers, plays an important role and participates extensively in all fronts. Indeed, religious educated leaders form the backbone of separatist politics in Patani in the 1970s.

Although the fronts have been weakened in recent years by factional conflicts and internal rifts, they have conducted occasional armed insurgence in the four Muslim provinces to remind the government as well as villagers of their existence. More significantly, all fronts assume that popular support is always potentially present, and, therefore, their relatively weak resources are concentrated more on external activities, to obtain international support and recognition. This external emphasis has further lessened local activities. None the less, the separatist movement persists. Its strength, like that of the Moros, lies in the fact that it is ethnically based, religiously motivated (see Aruri, 1977; Snow and Marshall, 1984), and led by a leadership group that dominates Patani society.

The Moro and the Malay Separatisms Compared

The Moro and the Malay separatist movements have more similarities than differences. The outbreak of the present separatist conflicts in Mindanao–Sulu and Patani are but the latest chapter in the centuries-old history of Muslim resistance against alien domination. The contemporary struggles were also triggered by the governments' acceleration of integration efforts and by the specific immediate cases of maltreatment that characterized the 'internal colonialism' of the existing socio-economic and political structure of the Muslim communities. To the Muslims, the central governments had proved to be insensitive to their demands and needs. The search for alternatives to secure the integrity of their communities became inescapable.

Both movements consist of loosely organized fronts whose strengths are less dependent on the principles of effective organization than on the fact that they are ethnically based, religiously motivated, and led by leadership groups that dominate their respective communities. The Moro fronts are led primarily by traditional aristocratic and secular-educated élites; the Malay fronts are led mainly by the religious élite which likewise dominates the leadership of Malay-Muslims in Patani society. In addition, the two struggles are supported, though to different degrees, by some sympathetic Muslim states. Such support has had significant effects on the separatist conflicts: the external supporters 'help to keep a conflict between a minority group and the majority-dominated government it challenges within the bounds of total defeat and total victory' (Suhrke and Noble, 1977: 209). More importantly, all fronts, except the communist-influenced MORO, are founded on Islamic ideology, though each has its own coloration. Islam has become the fronts' most powerful symbolic means of mobilizing support and of legitimizing sanctions for revolutionary action. Yet, despite the bond of Islam, both movements have become factionalized. Factional conflicts have been one of the major sources of past and present weaknesses of the Muslim struggles in confronting external threats. In the case of the Moros, ethno-cultural splits and family backgrounds have been major causes of factional conflicts. Although the Malays have no ethnic divisions, their conflicts derive primarily from differences in ideological inclinations and educational and family backgrounds. Disunity among the Muslim states has further strengthened factional battles in both movements.

Despite their differences, the fundamental concern of the Moros and the Malays is to seek their survival as Muslim peoples. They believe that the preservation of their Islamic communities requires freedom from domination by Christians and Buddhists, respectively, in matters that most impinge on their identity as Muslims. These include such matters as religious activities, family life, education, and jurisdiction over land and economic resources. However, there can be no freedom until there are changes in the structure of their relationships to their respective governments. To obtain these changes, both movements have devised a two pronged strategy: attracting international attention to their dilemmas as oppressed Muslim minorities and employing guerrilla-style wars against the central governments.

The fundamental difference between the two struggles is the intensity of the conflicts. The Moros have been able to stage a 'war of attrition' that not only stands as one of the Philippines' most serious internal conflicts, but also led the Marcos government to sign the Tripoli Agreement and, more recently, President Aquino to meet some of their leaders in an attempt to initiate a negotiation settlement. In contrast, the Malays have been able to achieve only a low level of violence in southern Thailand. There are several reasons. One is that the Thai government's policies of assimilation and suppression have, since the late 1960s, been restrained by anticipated strong reaction from the neighbouring Malaysian government. Secondly, the Malays are less willing or able to develop international contacts (Suhrke and Noble, 1977: 209–10). And, finally, the Malays lack charismatic and aggressive leaders capable of providing a serious challenge to the Bangkok government.

1. On 15 April 1980, the Sangguniang Pampook (Legislative Assembly) approved the Marawi City government's proposal to change the name 'Marawi City' to 'Islamic City of Marawi'.

2. The OIC meeting in Libya in 1972 adopted a resolution to investigate the 'plight of the Muslims of Southern Philippines' and created the Quadripartite Ministerial Commission comprising Libya, Saudi Arabia, Somalia, and Senegal. It was ratified by the Islamic Summit of 1973.

3. According to Inamullah Khan (1979: 14–17), the Moros under the MNLF leadership were fighting against the Philippine armed forces of about 200,000 men in June 1976. Between 1969 and 1976 the costs of war were considerable. In Cotabato, the death toll was estimated at about 20,000, the number of wounded at 8,000, and the number of displaced at 100,000. In Lanao provinces the corresponding figures were 10,000, 20,000, and 70,000; in Sulu and Tawi-Tawi, 10,000, 8,000, and 100,000; in Zambaoangas, 10,000, 10,000, and 40,000; and in Basilan 10,000, 8,000, and 40,000 (cf. Hussin, 1981: 257–60; cf. Miranda, 1985). There was hardly any Moro province that had not been affected adversely. The loss of property was estimated to have cost 300–500 million pesos.

4. The Islamic Solidarity Fund was one of the subsidiary organs of the Organization of the Islamic Conference. It was founded in 1974 to meet the needs of Islamic communities by providing emergency aid and the wherewithal to build mosques, Islamic centres, hospitals, and educational institutions.

5. When Sultan Rashid Lucman died, his cousin, Dr Yusoph A. Lucman, took over the BMILO leadership as Chairman with Jamil D. Yahya as its Secretary (BMILO, 1984a: 1; *Arab News*, 9 November 1984; *Friday Times*, 25 March 1985). Nurdin Lucman, a son of the late Sultan, is also an important BMILO leader. The three top leaders of the BMILO operate in three different bases: Yusoph in Cairo; Jamil in Mecca, and Nurdin in Marawi City.

6. Mistakes often occur in some of the government religious textbooks, perhaps due to the inexperience of Thai officials in the Islamic subjects. In 1980 the Ministry of Education distributed a religious textbook containing several drawn portraits of the Prophet Muhammad (S.A.W.), which is categorically prohibited in Islam (Office of the Committee on Education, 1980). Such a blunder on the part of the Ministry of Education resulted in many protests against the Thai government by Muslim leaders and Muslim governments.

7. Badri Hamdan, now in his early fifties, is a graduate of Al-Azhar University in Cairo and a former lecturer at Chulalongkorn University in Bangkok.

8. Barisan Nasional Pembebasan Patani (BNPP) was renamed in 1986 by its general assembly to Barisan Islam Pembebasan Patani (BIPP) to stress its Islamic inclination.

4
Leadership in the Moro and the Malay Societies

ÉLITE theorists believe that concentration of power in a small group of élite is inevitable in modern society (Michels, 1915; Mosca, 1939; Bottomore, 1964; Pareto, 1966; Olsen, 1970). The population of every modern society is differentiated according to status, wealth, and power. The majority of any given population may be thought of as the base of a steep pyramid. If social class is defined as the measure of one's position along the vertical slope of the pyramid, the higher the class-standing of any individual, the smaller the percentage of the population that belongs to his social class. In other words, the élite and the rich are few, while the commoner and the poor are many (Greene, 1974: 16). Despite their small numbers, the élite influences many aspects of social life and provides leadership in shaping the structure of society. 'Most simply,' notes Lasswell, 'the elite are the influential' (Lasswell and Lerner, 1965: 4). They include not only leaders or those who engage in sociopolitical activities, but also the strata of society from which they are generally drawn (Lasswell and Lerner, 1965: 4–12).

According to the fourteenth-century Muslim thinker, Ibn Taymiyyah, leadership or authority is indispensable in the social structure of human life. Prophet Muhammad (S.A.W.) had charted the pattern of even the smallest organization by instructing that one among three who may be travelling be designated leader (Makari, 1983: 135; Qamaruddin Khan, 1973). This illustrates that Islam affirms the necessity and the inevitability of a leadership structure in society.

Leadership in Moro Society

Despite decades under the rule of the Filipino Christian government, Moro society has remained unintegrated. It has retained most of its traditional features and institutions, though in modified form. With respect to leadership, there exist three categories of Moro élite: traditional, secular, and religious. These leadership groups play various roles in contemporary Moro society.

The Role of the Traditional, Secular, and Religious Élites in Moro Society

A social system consists of sets of patterned and related activities. These activities may be termed 'roles'. Roles are '(1) units of conduct which by their recurrence stand out as regularities and (2) which are oriented to the conduct of other actors' (Gerth and Mills, 1953: 13). An organization of roles, one or more of which is understood to serve the maintenance of the total sets of roles, is called an institution (Gerth and Mills, 1953). Political order comprises those institutions within which men attain, wield, or influence the distribution of power within social structures (Gerth and Mills, 1953: 26). Leadership, therefore, can be measured primarily in terms of roles which individuals or groups of individuals play in such institutions.

Members of the Moro traditional élite are the aristocrats: the sultans, the *datu*, their descendants, and those who attain aristocratic status by their own personal achievements (see Chapter 1). Traditionally, the pervasive influence of the sultans and the *datu* extended to virtually all spheres of life (Stewart, 1977: 66). They were also regarded by the villagers as charismatic leaders whose charismatic grace (*barakat*) adhered in them and could be passed on hereditarily (Kiefer, 1972a).[1]

Despite the imposition of the Western political structure on top of the traditional system, Moro social activities continue to revolve around the aristocrats (Glang, 1974: 282). Members of different ethno-cultural groups still recognize the existence of their respective royal houses. Even though the sultans and the *datu* no longer have temporal power, they remain an influential and powerful élite group within the Muslim community. Three basic factors contribute to the survival of the traditional leaders. First, they have been able to adapt themselves to the political machinery of the modern Philippine state. Their traditional ranks and roles were legitimized and institutionalized with the introduction of popular elections: with their influence and accumulation of wealth, they easily won in elections (Madale, 1984: 180). Secondly, the aristocratic institutions headed by sultans and *datu* have always been looked upon as representing the authority of Islam and *adat* (customary law) which constitute the major points of reference in terms of which the Moros define their society and their place within it (Kiefer, 1972b: 90–1). Thirdly, the traditional élite belongs to an institution that fortified the Moros through centuries of struggle against foreign domination (Glang, 1969: 56).

The second category of Moro leaders belongs to the secular élite. They are the professionals who '(a) perform an activity on the basis of specialized training and, usually, higher education; (b) work full-time for (c) a remuneration which is their main source of income' (Pusic, 1969: 108). They are people in a special social role, such as lawyers, doctors, engineers, businessmen, and lecturers, among others. The Moro professionals are the product of local and foreign secular colleges and universities.

TABLE 4.1
The Commission on National Integration Programme:
Number of Scholars and Graduates, 1958–1967

Year	Scholars	Graduates
1958	109	20
1959	460	60
1960	610	110
1961	620	110
1962	930	180
1963	1,020	130
1964	930	150
1965	960	170
1966	1,460	250
1967	1,210	211
Total	8,309	1,391

Source: Filipinas Foundation (1971: 163).

The secular élite gain leadership status not only through their special social role, but also through what Lynch terms the 'alliance system' (1959: 49–53). Alliances are forged between professionals, such as lawyers, and individuals of lower socio-economic status, such as ordinary villagers. For example, a powerful lawyer may, out of a conscious feeling of professional responsibility or of charity, provide assistance to a poor villager who needs his professional help. In such a case, a dyadic relationship may grow, with the villager becoming a dependent partner, reciprocating the lawyer's help through small gifts or by providing certain services (Arce, 1983: 59). While they are still comparatively small in number, members of the secular élite increasingly play an active leadership role in the Muslim community. Some of the Moro secular élite were the product of the Commission on National Integration College Scholarship Programme which provided college scholarships to members of minority groups. Table 4.1 shows the number of scholarships and graduates during the years 1958–67. Some of these graduates emerged as leaders of the separatist struggle, notably Nur Misuari and his associates.

The third group of Moro leaders is the religious élite. Members of the religious élite are men who engage in religious works and activities. They include three main groups: mosque personnel (officials), such as *imam, khatib,* and *bilal*; religious teachers, *ustaz (pendita* or *guru),* and preachers *(du-ah)*; and pious men *(ulama)*[2] and *hajji* (see Chapter 2).

It should be noted, however, that membership of these leadership groups is not mutually exclusive. Depending on the individual's qualifications, a member of one élite group can at the same time be classified as a member of the other élite groups. For example, a *datu* who is also a *hajji* and a businessman can be identified as a member of all three groups. On the other hand, a *hajji* or a businessman who is not a *datu* does not qualify to become a member of the traditional élite be-

TABLE 4.2
Moro Leadership Group Combinations

Leadership Group	Possible Leadership Group Combinations			
Traditional élite	T	T + S	T + R	T + S + R
Secular élite	S	S + R		
Religious élite	R	R + S		

Notes: T = Traditional élite
S = Secular élite
R = Religious élite

cause he is not an aristocrat. Table 4.2 indicates the possible leadership group combinations for a member of each élite group.

In the five provinces (Basilan, Lanao del Sur, Maguindanao, Sulu, and Tawi-Tawi) where Muslims constitute the majority, the Moro political and social activities revolve around the local governmental and religious institutions. There are three main levels of local government in the Moro provinces: regional, provincial, and municipal. As a measure of the extent to which each élite group participates in the local governmental institutions, 125 elected and appointed contemporary Moro leaders of various capacities—senators, congressmen, governors, mayors, assemblymen, and regional directors—are examined. By examining their family names, royal titles, and professional and religious qualifications, their respective leadership group classifications have been identified.

At regional level, Autonomous Governments in Central and Western Mindanao (Regions IX and XII) came into existence as a result of the Tripoli Agreement. Table 4.3 shows the distribution of leadership groups among the elected and appointed Muslim officials of Autonomous Region XII. The majority of the elected Muslim officials of Autonomous Region XII, as illustrated in Table 4.3, are members of the traditional élite. This suggests that the traditional leadership group is still influential among the Moro voters. One may argue, however, that the voters were

TABLE 4.3
Leadership Group Distribution among Muslim Officials,
Autonomous Region XII (per cent)

Leadership Group Classification	Assemblymen and Members of Executive Council	Regional Directors
T	62.5	28.6
S	18.8	61.9
T + S	12.5	9.5
T + R	6.2	–
Total number of leaders	16	21

Notes: T = Traditional élite
S = Secular élite
R = Religious élite

TABLE 4.4
Leadership Group Distribution among the Elected Muslim Leaders at the Provincial and Municipal Levels (per cent)

Leadership Group Classification	Senators and Congressmen	Governors and Vice-Governors	City and Municipal Mayors
T	57.1	52.9	45.6
S	14.3	29.4	21.1
R	–	–	14.0
T + S	14.3	11.8	3.5
T + R	–	5.9	15.8
T + S + R	14.3	–	–
Total number of leaders	14	17	57

Notes: T = Traditional élite
S = Secular élite
R = Religious élite

also influenced by the then government-controlled political party, Kilusang Bagong Lipunan (KBL) or New Society Movement. But those candidates who were supported by the KBL were usually already influential leaders who commanded support among the villagers. In fact, most of the traditional leaders who were elected were, in the words of Michael Mastura, 'predetermined leaders' (see George, 1980: 99). In the case of the appointed regional directors, the majority of them are members of the secular élite, for professional tasks require professionals. Nevertheless, almost one-third of the regional directors are members of the traditional élite. Thus, the traditional and secular élites play an important leadership role in governmental institutions at the regional level.

At the provincial and municipal levels, the distribution of leadership

TABLE 4.5
Leadership Group Distribution among the Elected and Appointed Muslim Leaders at the Regional, Provincial, and Municipal Levels (per cent)

Leadership Group Classification	Elected Leaders	Appointed Leaders	Elected and Appointed Leaders
T	50.0	34.8	47.2
S	21.6	56.5	28.0
R	7.8	–	6.4
T + S	7.8	8.7	8.0
T + R	10.8	–	8.8
T + S + R	2.0	–	1.6
Total number of leaders	102	23	125

Notes: T = Traditional élite
S = Secular élite
R = Religious élite

groups among the elected Muslim leaders is shown in Table 4.4. From this table, it is clear that a significant majority of the elected provincial and municipal Muslim leaders are members of the traditional élite. The secular élite, on the other hand, is in a relatively impressive second position, while the religious élite comprises only a small percentage of the municipal mayors. This again illustrates the dominant leadership role of the traditional élite at the grass-root as well as the provincial levels. Table 4.5 shows the distributions of leadership groups among the elected and appointed Muslim leaders at all three local levels. It is evident from the table that members of the traditional élite dominate the elected Muslim leadership at all levels of local government. They also represent the majority of the overall elected and appointed contemporary Muslim leaders.

In addition to governmental institutions, religious institutions, such as mosques, *madaris*, and Muslim associations and organizations, are important in Moro society. There are many hundreds of mosques in the Mindanao–Sulu region (Majul, 1977: 779–84; Madale, 1976: 12–14). Many, if not most, of these were erected through the efforts of the traditional élite, especially in areas where there is a profusion of sultans and *datu*, as in Lanao del Sur (interviews with several *ustaz* in Marawi City, 14 November 1984). This is not an unusual practice in a Muslim community, for in the early days of Islam, the building of a mosque was an obligation on the rulers, and it became a pious work, 'according to which the Prophet said: "for him who builds a mosque, God will build a home in Paradise"' (Gibb and Kramers, 1953: 335). Prominent leaders also built mosques which were the centres of their activities. Such traditions have been followed consciously or unconsciously by the Moro traditional leaders. Nevertheless, the mosque officials remain dominant functionaries. In addition, preachers and *ustaz*, who teach or are involved in other religious activities, are also regarded as important leaders in the affairs of the mosque.

The second group of religious institutions comprises religious schools or *madaris*. *Madaris* as parochial schools have been with the Moro community for centuries. As William Dampier, an English explorer who visited Mindanao in 1687, noted: '... in the city of Mindanao they speak two languages indifferently; their own Mindanao language, and the Malays.... They have schools, and instruct the children to read and write, and bring them up in Mahometan religion' (quoted in Mastura, 1982: 6; cf. Majul, 1970: 1). Indeed, *madaris* have been the Muslims' basic instrument in preserving and transmitting Islamic teachings, customs, traditions, and beliefs (Pahm, 1982: 44). The Moro villagers acquire their religious knowledge, particularly about *ibadat*, from *madaris* which offer classes not only in schools, but in mosques and private houses. During the American regime, the *madrasa* system was discouraged; it was not until after the Second World War that *madaris* re-emerged, as a result of the resurgence of Islam (Gowing, 1962: 57–65). In the early 1980s, there were about 1,137 *madaris* in the Moro area (Ministry of Muslim Affairs, 1983: 34). They function as non-formal schools aimed at educating both youth and elders in the knowledge of Islam, the Arabic language, and other subjects patterned after the curriculum taught in the Middle East (Sarahabil, 1982: 26; Isidro, 1979). Like the mosques, the *madaris* are looked up to as a symbol of Islam and are regarded as essential to the lives and culture of the Moros. As Mercado (1981: 52) suggests, the mosques and *madaris* are the only ones which remain the instruments whereby they can affirm their 'Moroness'.

Most of the *madaris* are organized and maintained by the united efforts of *ustaz* or preachers and local leaders such as *datu* (Abdul Ghafur Madki Alonto, 1982: 32; Sarahabil, 1982: 26). Full-fledged *madaris* and Islamic institutes which combine some subjects of government-approved standard elementary and secondary curricula are normally managed by well-known religious or traditional leaders. For example, the Mindanao Arabic School and the Jamiatul Philippine Al-Islami in Marawi City are run by Sheikh Ahmad Bashir, a religious teacher who spent many years in Mecca studying Islam, and by the Alontos, one of the oldest aristocratic families in Lanao del Sur. The Sulu Madrasa Al-Islamia was established by Major Berley Abubakar with Mohammad Taja Omar, a preacher, as its headmaster. Back in 1951 Congressman Datu Luminog Mangelen became the patron of the Madarasatul Rasheeda in Cotabato while Sheikh Omar Bajunaid was its first headmaster. The Motamar Islamiah in Bayang was founded by another religious teacher, Ustaz Kali sa Bayang Monir. The Maahad Marawi Al-Islamie is run by Sultan Rashid Sampaco. In brief, the religious and traditional élite groups play a major leadership role in *madaris*. Some of the *madaris* have managed to secure financial assistance from various Muslim governments (Mastura, 1982: 10).

A major problem facing *madaris* as seen by some Muslim educators and Philippine government authorities is that they generally do not prepare their graduates for higher education and professional training in the Philippine educational system. Only a few graduates can look forward to further education in Islamic institutes abroad; most graduates find themselves educationally unprepared for full participation in Philippine socio-economic and political life outside their immediate communities (Gowing, 1979: 71). To solve the problem, these educators and the government authorities, some of whom see *madaris* as a training ground for militant Muslims, feel that *madaris* should be integrated into the public school system under such schemes as the Fund for Assistance to Private Education (FAPE) and the Educational Development Implementing Projects Task Force (EDIPTAF). This means that the *madrasa* administrators must be prepared to accept the directives of the education authorities in the matter of accreditation, educational management, and administration (Mastura, 1982: 11). To some religious teachers who view education as a medium for Islamic *dakwah* (the call of faith), however, the introduction of the government's secular curriculum into *madaris*, as the Malay-Muslims in southern Thailand have experienced, would not lead to a constructive result. As Pitsuwan has observed, 'Since the introduction of secular education into its curriculum, the products of the *pondoks* have become mediocre and, in some cases, valueless' (quoted in Mastura, 1982: 11).

The last category of religious institutions comprises Muslim associations and organizations, such as the Ansar El-Islam, the Muslim Association of the Philippines, and the Agama Islam Society. They were formed for a mixture of religious and social as well as political purposes. The combination of such purposes was inherent in the character of Islam, which holds that religion and politics cannot be separated. Through their activities, the Muslim associations and organizations have helped enhance the sense of Islamic consciousness and of Muslim identity and solidarity across Moro ethnic–cultural group lines (Gowing, 1979: 187). The building of mosques and *madaris*, which is part of their efforts, for instance, strengthens the religious consciousness and solidarity of the Muslims.

The first all-Muslim organization, the Sarikatul Islam Association, was formed in Zamboanga in 1924. In the same decade, the Muslim Association of the Philippines (MAP) was organized by Indian Muslim businessmen in Manila. Later, it became the hub of Muslim affairs. Some of its achievements benefited Muslim education, including the establishment of *madaris* in the provinces and securing financial assistance and scholarships from Muslim governments such as Egypt, Pakistan, and Indonesia (Mastura, 1982: 10). In 1936 a group of Moro leaders in Marawi City formed the Kamilol Islam Society with the purpose of promoting Islamic education. The Society then founded the Madrasah Islamiah Kamilol Islam under the direction of Sheikh Muhammad Siddiq, Guru sa Marawi. Several other associations of similar purpose also emerged in different parts of the Moro areas. In 1969 the Ansar El-Islam was organized under the leadership of Senator Domocao Alonto. Its stated objective was 'the complete establishment of Islam in all aspects of their way of life' (Ansar El-Islam, 1974: 1). It struggled to achieve, among other things: 'a defined and guaranteed territorial jurisdiction; and a legislative power to adopt local laws based on the Holy Quran and the Sunnah of the Holy Prophet' (Ansar El-Islam, 1974). It was also an attempt to find a 'workable' solution to the demand for complete independence of Mindanao made by Datu Udtog Matalam's Muslim Independence Movement organized a year earlier. Other Muslim organizations include the Union of Islamic Force and Organizations, the Muslim Lawyers' League, the Muslim Progress Movement, the Sulu Islamic Congress, the Muslim Youth National Assembly, the Supreme Islamic Council of the Philippines, the Muslim Alliance of the Philippines, the Muslim Students' Association of the Philippines, and Sulu Muslim League. All are generally aimed at promoting the interest of the Moros and preserving the *ummah*. However, there are associations that are personalistic, forming and dissolving around particular groups of communal leaders in efforts to promote or preserve their socio-political positions (Noble, 1976: 407).

The extent to which different élite groups are involved within each organization is dependent upon its orientation. For instance, if the organization is oriented towards the promotion of Islamic education and consciousness, such as the Kamilol Islam Society, the religious and traditional élites play major roles. On the other hand, if the main purpose is to strengthen the sense of national identity and Muslim solidarity, as with the Ansar El-Islam, the traditional and secular élites are more extensively involved. In both cases, the traditional élite performs an important leadership role.

In the final analysis, the traditional élite is a dominant group in both local governmental and religious institutions. The group dominates many aspects of Moro social life and provides leadership in directing Moro society. The traditional élite is, in fact, still the most powerful and influential leadership group in the Moro community. As Gowing (1979: 55) wrote, 'Power in the present-day Moro society is largely in the hands of those who combine both traditional authority of their group and the constitutional authority of civil government' (cf. Abducal W. Alonto, 1983: 92). In the case of the religious élite, its leadership role is prevalent in religious institutions such as *madaris* and mosques. Its influence in the rural areas ranks second only to the traditional leaders. The religious élite play a valuable role in the attempts to keep the Islamic integrity of the community. Similarly, the secular élite has its own area of influence. However, because its number is still limited, particularly the non-aristocratic secular élite, the momentum of its influence and power remains restricted to certain areas only.

The Status of the Traditional, Secular, and Religious Élites in the Filipino-dominated State

Emerson (1960) once observed that the end of colonialism by itself removes only problems that arise directly from foreign rule; but the oppression of some groups over others continues. In the case of the Moros, their domination by the Christian Filipinos began as the American colonial control of the Archipelago ended. The structure of social relations between the two groups of peoples has been characterized as 'internal colonialism', the domination and exploitation of natives by natives.

The internal colonial system foments social dichotomy and deprives the minorities of their rights (Casanova, 1963; Hechter, 1975). The minorities are excluded from economic, social, and political opportunities and are held in low regard. They are usually the objects of contempt, hatred, and violence (Israeli, 1978: 13). In addition, the minorities tend to develop a set of attitudes and other behaviour that further differentiate them from the dominant group. For example, a significant majority of the Moros view themselves as 'Muslims', despite the prevalent feeling among Christians that this minority group considers itself to be 'Filipinos' (Filipinas Foundation, 1971: 116–17).

As shown in Chapters 1 and 2, the underprivileged status of the Moros takes various forms, ranging from socio-economic discrimination to not being treated decently. The Moro élites too are in a subordinate position in the Filipino-dominated state. The traditional élite, the sultans and the *datu*, is no longer recognized and given temporal power. In a Memorandum to the Secretary of the Interior in 1938, President Quezon gave the following instructions: 'These *datus* and sultans should never be allowed to have anything to do with functions that are official. They should be heard exactly and precisely as every citizen has the right to be heard on matters affecting the nation, his province, his municipality, or his district' (quoted in Gowing, 1979: 55). These instructions have become the guide-lines of Philippine government policy toward the traditional élite to this day. Even though the traditional élite found it expedient to seek public office in the Philippine political system, the openings were sparse. Many of the responsible offices in the Muslim provinces were reserved for Christians. The dissatisfaction toward the government's treatment of the traditional élite was expressed by Sultan Alaoya Alonto as follows: 'The Moro people want to set their house in order but how can they when the very key to their own house is not in their possession and perhaps the Moro may be locked in or locked out because the key to their own doors is not in their hands. This is indeed a sad tragedy' (quoted in Ralph B. Thomas, 1971: 269–70). In 1974 the Marcos government gave formal attention to the Moro traditional élite by acknowledging the existence of the 'Nineteen Royal Houses of Mindanao and Sulu'. However, many Muslims believe that this official attention to the Moro sultanates was part of the Marcos government's attempt to enlist the support of the Muslim aristocrats in its campaign against the Moro struggle. Nevertheless, this official attention to the Moro sultanates had two significant results. First, it drew the traditional élite closer to the government and thus made it more willing to get involved in government efforts to find a solution to Muslim problems. For example, barely a month after the sultans were invited to Malacanang Palace, the 'Federation of the Royal Houses of Mindanao and Sulu' sponsored the 'Muslim Conference on Government Policies and Programs for Muslim Mindanao', as discussed in Chapter 3. Secondly, it gave those younger-generation revolutionary leaders, such as Nur Misuari, Abdurasad Asani, Hatimil Hassan, and others who bid defiance to the traditional leadership, a basis for accusing many of the sultans and *datu* of being 'collaborators' of the Marcos regime.

Although certain members of the traditional leadership group have benefited from particular government policies which favoured appeasement of a few Muslim leaders, many of them, however, are no longer wealthy or powerful and distinguishable from the ordinary villagers, except by their honorific royal ranks. In fact, the traditional aristocratic élite, which once established an efficient government and contributed to the progress of the Moro people, is now reduced to a mere symbol of its past existence.

Like the traditional élite, the Moro secular élite is also dissatisfied with its situation. Despite their professional and educational qualifications, many members of this group find difficulty in securing a responsible job in the public or private sector. The main reason for this, apart from the nature of internal colonialism itself, is the preconception that the Moros cannot be trusted (Tongson, 1973: 20). Furthermore, very few of the non-aristocratic secular élite have occupied important elective political offices, such as governors and city mayors, because such offices are dominated either by Christians or by the traditional élite. Many of the secular élite, particularly businessmen and other professional groups, are constantly experiencing economic difficulties. Some have left the provinces in pursuit of employment opportunities in other Muslim countries such as Malaysia, Saudi Arabia, and Libya. Each year, a number of the Moro graduates of various universities in the Middle East remain abroad to obtain employment.

The status of the Muslim religious élite ranks last in the Filipino Christian society. This is not surprising because the Islamic religion has very little value in the view of the Christian Filipinos. In a letter to Senator Domocao Alonto, a Christian Filipino wrote, 'If the Muslims in the Philippines are poor and backward, it is because of their wrong religion and ideology, Islam.... You and your people should not compound your grievous historical mistake by clinging on to a religion that has only brought poverty, ignorance, and darkness to you and your communities' (Ahmad Domocao Alonto, 1979: 61). The religious élite is excluded from the mainstream of Philippine social mobility. Although members of the religious élite are among the educated in Moro society, their religious education or degrees are not recognized. Their occupational options are limited mainly to teaching posts in *madaris* and mosque functionaries. Being dependent on very low tuition fees and donations, the salaries of the religious élite are below subsistence level. Some religious teachers must teach in several *madaris* in different places to earn a living; many are unemployed.

The religious élite is isolated from the Philippine political arena. Its members have no political power and seldom stand for election to public office. The low political participation by the religious élite is due to many factors, including the language barrier, economic constraint, and limited knowledge of the Philippine political system. However, the main cause is that the Philippine government and the political parties do not make any efforts to recruit them into the national political orbit or to convince them that the Philippine political system could also serve their interests. On the contrary, the religious élite is perceived as representing an undesirable culture and a barrier to national development and progress.

In sum, Moro society has been transformed from the Islamic-oriented community of pre-colonial times to one of subordination under the rule of the Christian secular government. Accordingly, a sense of distress and alienation and of anticipated bad treatment has been part of the Muslim relationship with the ruling Filipinos. In this relationship, the inevitable outcome has been the struggle for liberation.

The Role of the Traditional, Secular, and Religious Élites in the Moro Liberation Struggle

Revolutionary leaders do not make a revolution. At best, they choose the means of revolutionary action and determine the movement's tactics and the timing of their implementation; the movement's ends and general direction are largely beyond their control. Part of the leaders' role, therefore, is to concentrate on issues and problems that affect various groups and sectors of the society (Greene, 1974: 26–7). Thus, the genius of Ho Chi Minh and Fidel Castro was their ability to mobilize broad support from among the diverse sectors of their respective populations. Similarly, the Moro's liberation struggle is the product of major events and problems that affected a large cross-section of the Muslim population.

Early in the struggle, the different leadership groups in the movement seemed to be dominated by members of the traditional élite. The top-echelon leaders, who held important positions such as chairmen and heads of committees, comprised mainly the traditional élite. On the other hand, many of the middle-level leaders, such as field commanders and officers, came from the secular élite group. Though members of the religious élite seldom appeared on the movement's top leadership list, they played an important leadership role in promoting the villagers' awareness of their role in the struggle as *mujahideen*. Table 4.6 lists some of the more active Moro leaders in the early underground movement, as discussed in Chapter 3.

Be that as it may, the composition of core leaders of the movement changed dramatically after the MNLF formed its official Central Committee in Tripoli in 1974. The Committee, known as the Committee of

TABLE 4.6
Some Active Moro Leaders in the Early Movement, 1970–1972

Name	Leadership Group Classification	Name	Leadership Group Classification
Udtog Matalam	T	Domocao Alonto	T + S + R
Salipada Pendatun	T + S	Rashid Lucman	T
Macapanton Abbas	T + S	Abul Khayr Alonto	T
Salamat Hashim	T + R	Yusoph Lucman	T + R
Udtog Matalam Jr.	T	Yahya Sheikh Moner	T
Nur Misuari	S	Saleh Loong	S
Abdul Baki Abubakar	S	Usman Imam	S
Abdurasad Asani	S	Ibn Al-Adil	S
Musib Buat	S	Farouk Carpizo	S
Abdul Karim Sidri	S		

Sources: MNLF (1982); George (1980).
Notes: T = Traditional élite
　　　　S = Secular élite
　　　　R = Religious élite

TABLE 4.7
Members of the Central Committee of the MNLF, 1974

Name	Leadership Group Classification
Nur Misuari	S
Salamat Hashim	T + R
Abdul Baki Abubakar	S
Giapur Ali	R*
Yusuf Abbas	R*
Abul Khayr Alonto	T
Hatimil Hassan	S
Abdurasad Asani	S
Abdul Hamid Lukman	S
Hajji Hassan Jamil	S + R
Ustaz Abijari	R
Usman Sali	S
Al-Caluang	S

Sources: George (1980); MNLF (1982).
Notes: T = Traditional élite
S = Secular élite
R = Religious élite
* = Students of theology or Islamic jurisprudence

Thirteen, was organized by Nur Misuari and Salamat Hashim with three other Moro students in Arab universities, namely Yusuf Abbas, Abdul Baki Abubakar, and Giapur Ali. Many members of the Central Committee were recruited from among Moro students who were studying in various Arab universities at the time. They were given specific portfolios in shadow-cabinet style (George, 1980: 227–31). The thirteen members of the Central Committee are named in Table 4.7.

As Table 4.7 indicates, the secular élite dominated the top leadership echelon of the MNLF Central Committee. The religious élite came second, whereas the traditional élite constituted only third place. Moreover, many members of the Philippine Student Union in Cairo and Moro students in different Arab countries gravitated toward the MNLF. These students were mostly potential members of the religious and secular élites. Thus, the MNLF, unlike the BMLO or MIM, was led mainly by young members of the secular élite who struggled not merely against Filipino Christian domination in order to preserve the *ummah*, but also to remove the domination of their own aristocratic class so that power relations within the Muslim community could be altered (George, 1980: 201–2; cf. Glang, 1969: 38).

Since the Tripoli Agreement, however, the MNLF has become factionalized. Each group has developed its own pattern of leadership. On the basis of the author's field observations (from October 1984 to April 1985), he is of the opinion that the leadership of Salamat Hashim's group has been influenced increasingly by the religious élite group. In 1984 it declared itself a separate organization known as Moro Islamic Liberation Front (MILF). On the other hand, the faction led by Dimas

Pundato (a member of the traditional élite) has moved closer towards the leadership of the traditional élite group, while Nur Misuari's group continues to be dominated by the secular leaders.

The role of the different leadership groups in the Moro liberation movement can thus be summarized as follows. The traditional élite played a crucial leadership role during the beginning of the struggle. Their leadership activities helped to promote unity among the Moros and to secure support from several Muslim countries. This was accomplished through the efforts of various organizations such as the UIFO, MIM, Ansar El-Islam, BMLO, and IDP. When the MNLF dominated the theatre of the Moro struggle, the top-echelon leaders changed from the traditional élite group to the secular élite. As the MNLF broke into three major factions, each of the three élite groups seems to have been attracted to a different faction.

Leadership in the Malay Society in Patani

The Malay society, like that of the Moros, has retained most of its traditional characteristics and institutions. Despite continuous efforts by the Thais to assimilate them, the Malays still revolve around their own Muslim leaders. These leaders may also be categorized under traditional, secular, and religious élites. Each élite group performs its respective role in the society.

The Role of the Traditional, Secular, and Religious Élites in the Malay Society

As discussed in Chapters 1 and 2, Muslim society in Patani is a close and well-knit community with much activity revolving around mosques and *pondok* (see Table 2.3). Nearly all aspects of villagers' affairs assume a religious significance and thus involve the religious élite.

The religious élite in Patani society can be divided into three categories: members of the Provincial Councils for Islamic Affairs (PCIA); members of the Councils for Mosque (CM); and religious teachers. There are 15 elected members for each of the PCIA, totalling 60 in the four Muslim provinces. They are elected for life by *imam*, who are heads of the CM. In addition, there are 2 *qadi* (Islamic judges) in each province who are also elected by *imam* and hold office until the age of 60. Thus, there are 68 such members of the Malay religious élite in the region.

Seven to fifteen members are elected for each Council for Mosque by adult Muslims in the communities. The head of the CM (the *imam*) and two of his deputies (the *khatib* and *bilal*) fill the lifetime positions while the other regular members serve a four-year term. There are 15,660 members of various CM in the Muslim provinces.

The third category of the religious élite comprises religious teachers. They are *tok guru* and *ustaz* who perform the role of religious educators at *pondok*, mosques, and *balaisa* (prayer-houses) in the Malay community. As indicated in Table 2.3, there are 1,215 *tok guru* and *ustaz* teaching in 202 Islamic schools in the Patani area.

The religious élite plays a leading role in most community activities, ranging from prayers to festivals. In most villages, the *imam*, *khatib*, and *bilal* lead the Muslim villagers in their daily prayers in mosques and *balaisa*. Important religious occasions such as *hari raya* (Muslim festive day), Maulud (occasion to commemorate the birth of the Prophet) and other ceremonies to mark special events such as *kenduri* (communal feasts) on the occasions of marriage, birth, and death always involve religious leaders.

Aside from the above functions, some of the religious élite who obtained degrees from Arab universities also work as *du-ah*. There are about 80 *du-ah* who receive monthly salaries and allowances from different Islamic call centres in the Middle East (interview with several *du-ah* in Narathiwat, 21 May 1985). The religious élite also act as the highest level of mediation within the community (Fraser, 1960: 164). Its members function as intermediaries between the people and the government. They are often called upon to present community matters to government authorities. Furthermore, they are normally active organizers of many community activities. As a result, religious leaders are treated with respect and are given positions of status and power in the society.

The most important group among the religious élite are religious teachers. They gain élite status because of their religious knowledge and teaching. Religious teachers are differentiated into *tok guru* and *ustaz*. Most *tok guru* are educated in the local *pondok* and mosques and in Saudi Arabia, Indonesia, and Malaysia. *Tok guru* and *ustaz* are considered the most authoritative source in matters pertaining to Islam. In addition, a few of them have the reputation of being *berkat* (blessed) and, many villagers believe, possess mystical powers. Hence, they also play the role of mystics to whom villagers resort for various types of assistance and cures. It should be noted, however, that the religious élite, especially religious teachers, are not an economic élite. The roles of the religious élite do not include that of employer, landlord, or middleman. Muslim villagers do not expect the religious élite to play economic roles. Those who do are often viewed as greedy and selfish. Because of this, *tok guru* and *ustaz* become closely related to the people and accordingly are highly respected and influential.

The Provincial Council for Islamic Affairs is a government Islamic institution at provincial level under the Department of Local Administration, Ministry of the Interior, and the Department of Religious Affairs, Ministry of Education. In theory, members of the PCIA are 'learned men' in different fields of Islam. They are elected to advise the provincial governors in matters concerning Islam and the welfare of the Muslims (Omar Farouk, 1981: 113). In practice, however, not all members of the PCIA are learned men, religiously or otherwise. It is estimated that not more than 30 per cent of the PCIA members are capable of giving advice to the Thai authorities on matters regarding Islam (Che

Man, 1983: 87). In fact, many Malay-Muslims regard members of the PCIA in the four provinces as 'second-rate' religious leaders and rarely seek their religious advice and *fatwa*. They also feel that religious advice given by the PCIA may not be completely free of the influence of the Thai authorities.

Similarly, the Islamic judges, who assist and advise civil judges in cases involving Muslims in matters regarding marriage and inheritance, are not always turned to. This is due to the fact that disputes involving family and inheritance can be settled by *imam* or by religious teachers. Besides, bringing the case to court would require both plaintiff and defendant to pay costly lawyers' fees.

The Council for Mosque is also a government Islamic institution which is directly under the authority of the PCIA. As members of the CM, the *imam*, *khatib*, and *bilal* are leaders in prayers and other religious rituals. They command respect from members of the community. If they are persons with additional leadership qualifications, such as *tok guru*, *ustaz* or *du-ah*, they are highly respected and influential. The remaining regular members of the CM are not necessarily religious individuals; they are men of different backgrounds who gain CM membership through their merits, reputations, and activities within the community. As members of the CM, they work with the *imam* and his assistants in matters pertaining to activities organized by the mosque. Unlike members of the PCIA, they are more in tune with the villagers' interests and the welfare of the community.

The secular élite in Malay-Muslim society comprises mainly local Muslim government functionaries, such as schoolteachers (*khru*), commune headmen, and village headmen. These low-ranking local civil servants gain élite status because they are members of the local government bureaucracy and earn comparatively substantial incomes. They are Malays with Thai educational backgrounds to whom the villagers, due to their language barrier, resort for help and advice in secular and government matters. Over the 1980s the influence of Muslim local government officials has increased substantially due to their increased participation in community and religious activities, while the reputation of religious leaders has dropped because the Thai government has been able to increase its control over two important Malay-Muslim institutions—*pondok* and mosque. Nevertheless, the local Muslim government servants do not command as much influence as the religious élite do, for they are seen by some Malays as government servants whose main duties are to protect government interests. A small group of Malay businessmen who play the roles of middlemen, traders, and contractors can also be regarded as part of the secular élite due to their professionalism and well-to-do status. This small group, however, is unable to lead the villagers because they lack prestige in a society where money-making does not provide high status.

The third leadership group is the Patani-Malay traditional aristocratic élite. Its members are the descendants of Malay *raja* or *chaomuang* who were appointed by the Thai government as rulers of the seven Malay provinces of the conquered and divided Patani kingdom. During the era of the seven provinces (1816–1906), there existed a total of 29 Malay *raja*: 7 in Patani; 5 in Yaring; 4 in Raman; 4 in Ra-ngae; 4 in Yala; 3 in Saiburi; and 2 in Nhongchik (Bangnara, 1976: 68–80; Bunnag, 1971; Ibrahim Shukri, n.d.). They were the ancestors of the contemporary Patani aristocrats.

After the Malay *raja* were replaced by Thai governors and the traditional system with the Thai secular political structure at the beginning of the twentieth century, the Malay aristocrats remained the dominant political force in Patani society. They led most of the earlier resistance because they wanted to regain their power. However, they failed to recapture political authority and lost much of their wealth, and many of the royal families left the region to seek refuge with their relatives in neighbouring states of Malaya. In other words, the Patani-Muslim aristocrats lost both their political and economic bases. They are no longer as powerful and wealthy as they used to be. In addition, some of the royal houses have given their loyalty to the Thai government. For example, the royal families of Yaring District in Patani Province are well-known as loyal supporters of Bangkok—which is not surprising because their ancestors were appointed rulers by Bangkok. As a result, the Malay traditional aristocratic élite is no longer perceived by many Malay-Muslims to be their legitimate leaders. More importantly, the aristocrats themselves have no intimate relationship with the common people. Most of them view themselves as of royal blood and feel that they should behave and be treated differently. This attitude creates a gap between them and the villagers, and thus diminishes their influence. Writers such as Pareto and Mosca have observed that the status of élites will decline if they fail to identify themselves with the interests of the masses (Bottomore, 1964).

Thus, while the secular élite, the local Muslim government functionaries, are respected for their association with state authority and while the traditional aristocratic élite, the descendants of the deposed Patani-Malay royal houses, are seen as representing the Malays of the glorious past, the most honoured and influential among the Malay-Muslims are men of religious reverence such as *tok guru*, *ustaz*, and *imam*. They are sources of religious education and spiritual guidance. The villagers, young and old, regard themselves, with varying degrees of explicitness, as the students or followers of these religious leaders. After all, the ultimate objective in life is to become *orang beriman* (a pious man). As in the past, religion and religious leaders dominate present-day Patani society. They constitute an important element which perpetuates the social isolation of the Malay-Muslims.

The Status of the Traditional, Secular, and Religious Élites in the Thai-dominated State

The Patani region was conquered and incorporated into the Thai nation-state. The structure of social relations between the ruling Thais and the ruled Malays, like that of the Moros, is based on domination and exploitation. All the phenomena found in internal colonial societies exist in Patani. The economy of the region is in an unsatisfactory condition compared to other sections of the nation. Most of the wealth in the provinces is in the possession not of the Muslims but of the Thais and Chinese. The Malays are the rural-dwellers and are at a level little better than subsistence. Furthermore, they are viewed with the typical colonial stereotype: Malay-Muslims are 'unreasonable' and 'good for nothing'.

Though the Thai government has established government Islamic institutions, such as the Provincial Council for Islamic Affairs, the Council for Mosque, and the office of Islamic judge in the four provinces, these institutions, especially the PCIA, do not represent the interests of the Malays. For many Muslims, they are essentially instruments of control and means of disseminating Thai government policies.

In the context of the larger Thai society, the different Malay élite groups are also subordinate in almost every aspect. They are excluded from the mainstream of Thai social mobility. Although the traditional and secular élites are mostly Thai-educated, they have never been appointed to responsible government positions such as Governor (except for the late Governor Termsak Semantarat, an aristocrat from Satun, who was appointed Governor to appease the Malay people after the so-called Patani Massacre in 1975). Of 28 district officers in the four Muslim provinces, only 2 are Malay-Muslims. In the case of the religious élite, their religious education and degrees are not recognized. They are unable to gain employment in the Thai system; their occupational options are limited to teaching in *pondok*.

In terms of political power, all Muslim élite groups are isolated from the political arena and have very little power in Thai politics. Religious leaders seldom stand for political elections, locally or nationally. There are many reasons for this, ranging from the language barrier to limited economic capacity. The real cause, however, is that while members of the religious élite do not believe that participation in the Thai political system could serve their interests, the government and Thai political parties are not interested in recruiting them into the political orbit either. On the contrary, the Thai government regards religious leaders, especially religious teachers, as a source of political resistance and treats them all as potential rebels. Hitchner and Levine (1973: 80) also suggest that the traditional élite (including the religious élite) in modernizing countries has tended to support regionalism against national unity.

In contrast to religious leaders, secular and traditional leaders are active in Thai politics, particularly since the birth of a short-lived democracy in Thailand in 1973. For example, of 9 national parliamentary seats contested in the four provinces in July 1986, 7 were won by Muslims. Table 4.8 lists Members of Parliament in the four Muslim provinces following the 1986 general election. However, these Muslim congressmen gain no real power in Thai politics, for they hold only 7 seats out of 324. Moreover, they are not considered 'trusted' members of their respective Thai political parties.

TABLE 4.8
Members of Parliament in the Four Muslim Provinces of Southern Thailand, 1986

Name	Province	Religious Affiliation	Leadership group Classification
Den To'mina	Patani	Muslim	S
Sudin Phujutthanon	Patani	Muslim	S
Preecha Bunmi	Patani	Buddhist	–
Say-nee Madakakul	Narathiwat	Muslim	S
Areephen Uttrasin	Narathiwat	Muslim	S
Phibun Phongthanet	Narathiwat	Buddhist	–
Adul Phuminarong	Yala	Muslim	S
Wan Mohammad Noor Mattha	Yala	Muslim	S
Chirayut Nao-waket	Satun	Muslim	S

Source: Town Hall of Narathiwat Province, Narathiwat, Thailand, 29 July 1986.
Note: S = Secular élite

The Muslim élites are also discriminated against culturally. They are perceived by many Thais as representing a foreign and undesirable culture. This is because the religious élite symbolizes Malay identity and is well known as a stronghold of cultural resistance. The Muslim élites are aware of their subordinate status in the Thai-dominated state. Some of them feel that it is their religious duty to strive for better alternatives, for the Holy Quran (XII: 11) admonishes, 'Verily never will God change the condition of a people until they change it themselves.'

The Role of the Traditional, Secular, and Religious Élites in the Malay Liberation Struggle

As discussed in Chapter 2, the early Muslim resistance to Thai domination was characterized by uprisings led by the deposed indigenous rulers of the seven Malay provinces in their attempts to regain their power. As the effect of Thai control over Malay society increased during the reign of King Vajiravudh, religious leaders like To' Tae and Hajji Bula, who feared their religion and culture would be stamped out, joined the aristocrats in resisting Thai rule. However, the leadership of the resistance movements at this stage was still confined to the former Malay ruling élite and their relatives.

The period under the ultra-nationalist government of Phibun (1938–1944) had been full of political protests and resistance that raised the political consciousness and sharpened the skills of the religious élite in mass mobilization. 'Slowly,' writes Pitsuwan (1982: 115), 'the *ulama* had been transformed from the role of "power brokers" to political activists.' This, in a way, was a matter of filling a power vacuum left vacant by the former *raja* and aristocrats who, by the end of the Second World War, were exhausted and were in a state of political incapacity due to Thai government antagonism. Many of them moved to Malaya to continue their struggle there.

While GAMPAR was formed in Kelantan by a son of a former Patani *raja*, members of the religious élite headed by Hajji Sulong and some secular nationalist leaders like Wae Semae Muhammad and Wae Useng Wae Deng organized the Patani People's Movement that turned the religious leaders into 'active seekers of power, which was deemed necessary for shaping and moulding of the community' (Pitsuwan, 1982). It was the first time that the leadership of the Muslim resistance movement was dominated by the religious élite group. The PPM in turn stimulated the outbreak in Kampung Belukar Samok in 1947 and the Dusun Nyor Revolt in the following year. The unrest was instigated by religious leaders such as Hajji Abdul Rahman (To' Paerak) and Hajji Mat Karang, and secular nationalist leaders such as Che Senik Wan Mat Seng and Zakaria Lalo. Along with this transition of leadership from the traditional aristocratic élite to the religious and secular leaders, the liberation movement also transformed itself from one whose base was restricted (comprising mainly members of the ruling élite whose main objective was to restore their authority) to one with a broad base of popular support aimed at either irredentism with the Federation of Malaya or independence for the Patani people.

When the formal liberation fronts—the BNPP, BRN, and PULO—were organized at the end of the 1950s and the beginning of the 1960s, the religious and secular élite groups played major leadership roles, despite the fact that the chairmanship of the BNPP and PULO still remained in the hands of the aristocrats. In 1977 Tengku Abdul Jalal, Chairman of the BNPP, died and was replaced by Badri Hamdan, a member of the religious élite. In 1984 Tengku Bira Kotanila (Kabir Abdul Rahman) was forced to resign as Chairman of the PULO; 'Ustaz Abdul Hadi', a religious leader, assumed its chairmanship.

The author's investigation of the BNPP's leadership in 1981 revealed that there were 37 top leaders who held forty-one different positions in the Front's Central Working Committee (CWC), Working Committee (WC) of the Mecca branch, and Heads of Province. Of these leaders, 62.2 per cent were members of the religious élite; 29.7 per cent members of the secular élite; 5.4 per cent the traditional élite; and 2.7 per cent others. In addition, there were 44 Muslim students who were members of five different Working Committees of the BNPP's overseas branches. Thirty-one of them pursued religious education or were potential religious élite (BNPP, 1981b: 14–15; Che Man, 1983: 163). In 1986, 13 of the 17-member Working Committee of BNPP's Mecca branch were from the religious élite; the rest were secular leaders (interview with Amin Hamdi, Chairman of the Working Committee of BNPP's Mecca branch, Mecca, Saudi Arabia, 15 April 1986).

If one examines the holders of the most powerful positions in the BNPP, such as Chairman, Deputy Chairman, Secretary, section heads of the CWC and of the WC of the Mecca branch, Heads of Province, members of the Military Committee, and commander of the National

Liberation Army of the Patani People (NLAPP), it will be found that the distribution of these portfolios was as follows:

Religious élite	61.2 per cent
Secular élite	24.5 per cent
Traditional élite	6.1 per cent
Others	8.2 per cent

The dominant leadership role of the religious élite group is also true for the remaining fronts. This finding is consistent with the assessment of the Thai authorities who have always regarded religious leaders as among the Muslim 'anti-government' activists. Indeed, the religious élite forms the backbone of the present-day Malay liberation struggle and are 'keepers of the community's trust in the fulfillment of its political aspirations' (Pitsuwan, 1982: 115–16).

Leadership in the Moro and the Malay Societies Compared

Among the various minority groups in South-East Asia, the Moros in the southern Philippines and the Malays in southern Thailand seem to have the most in common. They share similar leadership structures. However, the two societies are dominated by different categories of élite group: the former by a traditional élite; the latter by a religious élite. This difference can be explained by the fact that the religious élite in Patani society rose to fill the roles left vacant by the *raja* and their ruling aristocrats who were deposed and dispossessed by Thai authorities. Moreover, Thailand provided no political machinery to enable the Malay aristocrats to adapt to the new political system. By the time Thailand held its first national election in 1933, the former Malay ruling élite was already in a state of political incapacity. On the other hand, members of the traditional élite in the Philippines found it expedient to seek public office in the Philippine political system. Perhaps, the rapidly diminishing influence of the former Patani *raja* was also the result of the fact that they became rulers by virtue of Bangkok appointment. Their family trees were not identical to those of dynasties who reigned in the Patani kingdom in the past.

Another contrast is that the influence of the Moro religious leadership in the southern Philippines is increasing while the authority of the Malay religious élite in southern Thailand is declining. This is because the power base of the latter, *pondok* and mosque, was tempered by the government's successful attempt to increase its control over these two Malay institutions through the General Education Department and government Islamic institutions. In contrast, *madaris*, mosques, and Muslim associations in the Philippines are still comparatively free from the control of the government.

The vigour of the secular élite group in the two societies is also slightly different, despite the fact that it has been gaining momentum in both societies. For instance, the Malay secular leaders are mainly low-ranking local government functionaries, whereas the Moro secular élite includes middle-rank government officials such as ambassadors and commissioners. Additionally, the Moro secular élite is able to use various Muslim associations and organizations to extend its boundaries of influence. The Malay secular leaders have been unable to do so due to the strict control of the Thai government over such activities.

One major parallel between these different leadership groups is that they are in varying degrees treated as underprivileged minorities rather than as privileged élites in their respective countries; the religious élites, especially, are seen as obstacles to social development and progress.

The most significant similarity between the Moro and the Malay peoples is that most of the élite groups are to a certain extent involved in the respective liberation struggles. However, the dominant leadership group is different. The Moros have been led by various élite groups since the movement divided into four major factions: the Nur Misuari Faction is dominated by the secular élite, and the Dimas Pundato group by the aristocratic leaders. The Salamat Hashim and Yusoph Lucman groups are led by the traditional and religious élites. In the case of the Malays, all four major fronts seem to be dominated by the religious élite. Nevertheless, all are motivated to different degrees by the desire to recover and preserve those aspects of their identity and way of life as *ummah* which they believe they have lost in the process of nation-building and integration. This desire has been expressed through the liberation fronts.

---

1. Max Weber described charisma as 'a certain quality of an individual personality by virtue of which he is set apart from the ordinary men and treated as endowed with supernatural, superhuman, or at least specifically exceptional powers or qualities' (quoted in Smelser, 1963: 355).

2. *Alim* (learned man) is an honorary title given to a person knowledgeable in religious matters. The *ulama* are active participants in the village religious activities and ritual ceremonies. They do their best to maintain the Islamic integrity of their communities (Majul, n.d.: 35–6). An *alim* may not have a formal religious educational background, but his knowledge, ritual experience, and piety command respect from the villagers. Since the criteria of measuring *alim* are relative and informal, it is difficult to estimate how many *ulama* are in the Moro community.

5
External Influences on and Government Responses to the Muslim Separatist Conflicts

ETHNIC separatist conflicts have a tendency to attract external involvement. According to Suhrke and Noble (1977: 3–5), an internal conflict that explicitly raises the question of national self-determination stimulates international response. 'Like revolution ... ethnic conflicts raise an ideological issue with international implications.' Secondly, while separatist activities are intermittent, the conflict persists. This persistent quality provides opportunity for external parties to interfere for their own purposes. Thirdly, separatist conflict is sometimes vicious and intense. It is likely to involve a degree of violence that attracts outside attention and demands that 'something be done' about the situation. Finally, because ethnic identities seldom coincide completely with state boundaries, separatist conflict in one state has implications in other states where there are ethnic kin. It motivates external kin to become involved (Suhrke and Noble, 1977: 5–7).

The Philippines

External Influences

In the case of the Moros, external involvement began with the resurgence of Islam and the rise of nationalism in the Muslim world after the Second World War. As discussed in Chapter 2, this resulted in a deepening of the Moro Islamic consciousness and strengthened Moro nationalist sentiment. This in turn enhanced the sense of Muslim identity across the Moro ethno-cultural groups. Hence, the Moros emerged as a strong political force.

More importantly, the involvement of Muslim states such as Malaysia and Libya enabled the Moros to elevate the level of conflict from fighting for equality and justice to a war of liberation demanding self-determination. Malaysia's involvement at the beginning of the conflict served to lay a foundation for the Moro separatist movement. As mentioned earlier, Malaysia trained and provided logistic support to Moro fighters. The support was not so much based on Malaysia's commitment to religious duty, as on reaction to the Philippine secret military plan against Sabah (see Chapter 3). As the *Straits Times* (11 March 1974) reported:

The Malaysians were extending support to the rebels to pressure the Philippine government to drop its claim on Sabah. Starting with a total of 90 men in five batches in early 1969 Malaysia provided guerrilla training to Muslims from Mindanao and Sulu.... Muslim rebels in the Philippines received from Malaysian sources at least 200,000 rounds of ammunition and 5,407 weapons ranging from hand grenades to machine-guns, anti-aircraft guns and a 52-inch tube-like device firing ammunition 30 inches long.

Though the Malaysian government later began to move towards co-operation in the Association of Southeast Asian Nations (ASEAN), Tunku Abdul Rahman continued to play an important role in promoting international support for the Moro cause. As Secretary-General of the OIC, in 1972 he endorsed the Moro case submitted to him and asked King Faisal of Saudi Arabia and President Qadhafi of Libya to help in persuading other OIC member states to support it (MNLF, 1982: 7). For the first time, the case of the Muslims in the Philippines was taken up by the OIC.

At the state level, Tun Datu Mustapha Harun, Chief Minister of Sabah (1968–1976), is claimed to have played a dominant role in assisting the Moro struggle. During 1972–6 Tun Datu Mustapha allegedly allowed Sabah to be used as a training camp, supply depot, communications centre, and sanctuary (Noble, 1983: 46). The Philippine government accused Sabah of allowing the Moro rebels to acquire about 100 motor boats in Sabah to be used to smuggle arms and ammunition to the rebels in Mindanao and to take rebel casualties back to Sabah for treatment (*Sunday Mail*, 23 December 1979). Furthermore, Sabah became a place of refuge for the displaced Moros. In 1983 it was estimated that between 160,000 and 200,000 evacuees were living in Sabah (Tunku Shamsul Bahrin and Rachagan, 1984: 196; *Straits Times*, 31 October 1983).

Several explanations have been given as to why Tun Datu Mustapha might have played an active role in accommodating the Moro struggle. First, he was committed to the cause and propagation of Islam. Under his leadership, important Islamic institutions such as the Majlis Ugama Islam (Islamic Religious Council), Office of Mufti (Office of Chief Judge), and United Sabah Islamic Association were created. More importantly, he belived that it was a duty of individual Muslims and Muslim states to help oppressed fellow Muslims, as enjoined by the Quran (IV: 75). The acceptance of large numbers of Moros in Sabah by Tun Datu Mustapha was also determined by an economic factor. Sabah was facing severe manpower and labour problems. The arrival of the Moro trainees and refugees helped to ease these problems. Finally, Tun Datu Mustapha allegedly had his own political antipathy towards the Philippine government. He claimed paternal lineage from the Sultan of Sulu and was always against Manila's treatment of the Moro people. He was also alienated by the Philippine claim to Sabah.

However, Tun Datu Mustapha's United Sabah National Organization (USNO) was defeated in the April 1976 election. This was partly because Sabahans, especially non-Muslims, felt that the presence of large numbers of Moros in Sabah, as a result of Tun Datu Mustapha's policy of accommodation, had adverse political and social implications. The predominantly non-Muslim Kadazans feared that the influx of the Moros would jeopardize their political and cultural status in Sabah (Tunku Shamsul Bahrin and Rachagan, 1984: 210). They perceived the tolerance shown to the Moros as part of a policy aimed at restructuring the communal balance in Malaysia. In addition, Philippines–Malaysia relations (cf. Noble, 1975) had suffered because of the 'Moro problem'; careful avoidance of such conflictual issues was needed to sustain the apparent calm in bilateral relations and the unity of ASEAN (Tunku Shamsul Bahrin and Rachagan, 1984: 210–11). Tun Muhammad Fuad Stephens, Tun Datu Mustapha's successor, therefore indicated that he would not follow his predecessor's policy with respect to the Moros. Datuk Harris Salleh, Tun Stephen's successor, and the present Chief Minister, Datuk Joseph Pairin Kitingan, have likewise not continued Tun Datu Mustapha's policy of accommodation. Nevertheless, the Moro separatists remain active in Sabah, and some groups such as the Nur Misuari Faction are still using Sabah as the base for their operations. The Malaysian government has never publicly admitted its involvement in the Moro struggle. No one, except those who were actually involved, could evaluate the amount of assistance given. But one thing is certain: Malaysian assistance gave the essential incentive to the Moro separatists and exposed their issue to the international community.

The second Muslim state to assist the Moro separatist movement was Libya. Libya's involvement was motivated partly by the dictates of Islamic brotherhood and the Quranic obligation to relieve the persecution of the *ummah*. Libyan assistance began in 1971 when its Information and Foreign Minister, Saleh Bouyasser, came to the Philippines during the World Assembly of University Presidents and met the Moro leaders to discuss Moro problems amidst increasing reports of massacre in the Muslim areas. According to separatist sources, Libya donated US$1 million to cover the expenses of 300 Moro recruits who were undergoing guerrilla training in Malaysia during 1971–2. After a meeting in Tripoli between President Qadhafi and Sultan Rashid Lucman, Domocao Alonto, and Salipada Pendatun, Libya agreed to provide assistance to the Moro struggle. When martial law was declared in September 1972, Libya started to deliver funds, weapons, and other equipment to the MNLF under Nur Misuari's leadership. Tun Datu Mustapha is said to have allowed Sabah to be used for transactions and as a supply depot. From 1972 to 1975, weapons and other defence supplies worth millions of dollars were delivered to the MNLF camp in Sabah. About US$35 million was contributed to the MNLF by Libya and the OIC during this period (Lucman, 1982: 5; Philippines, Ministry of Foreign Affairs, 1980a: 8–9). In addition, Libya made vigorous attempts, though without success, to exert its influence upon the OIC member states to impose economic sanctions against the Philippines. Despite the fact that the impact of Libya's support was only moderate, because of Qadhafi's limited popularity among the Muslim heads of state, it was significant in that it kept the struggle alive and motivated members of the OIC to deal with the MNLF (Noble, 1983: 47). Subsequently, falling into line with the position of other Muslim heads of state, Qadhafi's stance was changed from total backing of the MNLF to exerting pressure on both sides to make concessions. The shift was not solely to take heed of the OIC preference for a peaceful solution to the Moro problem, but reflected a realization that the Moro war against the Marcos regime was stalemated. The move coincided with the downfall of Tun Datu Mustapha in Sabah. The change in Libya's position led to the meetings in Tripoli between the Philippine government and the MNLF panels which resulted in the Tripoli Agreement.

Libya continued to play the role of mediator in an attempt to 'put an end ... to the war and leave no chance to those who are opposing the reaching of a solution' (Qadhafi, 1977). Even after the collapse of negotiations on details of the Tripoli Agreement in April 1977, Qadhafi still expressed his hope for peace. To this end, Libya refrained from campaigning for economic sanctions against the Philippines when the Marcos government appeared to begin to implement the terms of the Tripoli Agreement. However, Libya insisted on maintaining its assistance to the Moro struggle in order to pressure Manila to fulfil the obligation of the Tripoli accord. This was evident from the communiqué of the ICFM in Tripoli in May 1977 which called on the OIC countries to support the MNLF and to continue their efforts to mediate between the two parties. When the Aquino government replaced the Marcos regime in February 1986 and expressed its willingness to negotiate with the Moro separatists, the Muslim states, especially Libya, welcomed the gesture and promised to help persuade the MNLF to come to the negotiating table (*Far Eastern Economic Review*, 11 September 1986). As a result, a meeting between President Aquino and Nur Misuari took place in September 1986 and an 'autonomy agreement' was signed in Jeddah between Misuari and the government's negotiator, Aquilino Pimentel, in January 1987 (*Far Eastern Economic Review*, 15 January 1987).

Another Muslim state that has played a significant role in efforts to bring about settlement of the Moro problem is Saudi Arabia. It has been an active participant in the OIC Ministerial Committee of Four (Quadripartite Ministerial Commission) that has played a mediating role in the conflict since 1973. Through agencies such as the Muslim World League and Darul Ifta, Saudi Arabia has provided funds and sanctuary to different Moro fronts, including the MILF and BMILO. These agencies have also given assistance to a variety of projects in Moroland. In 1980 Saudi Arabia temporarily stopped its oil supply to the Philippines because of the Marcos government's failure to implement the Tripoli Agreement in good faith. Since 40 per cent of Philippine oil imports are from Saudi Arabia, such action matters (Noble, 1983: 48; Wurfel, 1985: 223). Iran, another supporter of the Moro struggle, cut off its oil shipments entirely, though only for a short period.

Unlike Malaysia, Indonesia did not support the Moro separatist movements against the Marcos regime, though it tried to put a stop to the conflict. Indonesian leaders have been concerned with their own internal and regional stability. Their hesitation to support the Moro rebels was largely a result of their own experience with militant movements in the 1950s and 1960s. Furthermore, Indonesia does not consider itself a Muslim state in the sense that Islam is not constitutionally regarded as a religion that provides the ruling principles for state policy. Hence, though it is a member of the OIC, Indonesia's foreign policies are not dictated by feelings of Islamic brotherhood and Quranic obligation.

In the interests of regional stability, Indonesia played a role as mediator between Malaysia and the Philippines. At the height of the Moro conflict, it attempted to initiate a deal by which Manila would renounce its claim to Sabah in exchange for Malaysia's agreement to suppress Tun Datu Mustapha. In 1974 Tun Datu Mustapha was reportedly warned by Jakarta's leaders about his co-operation with Qadhafi, who was calling for sanctions against the Philippines (Suhrke and Noble, 1977: 190). But Indonesia's efforts to persuade the concerned parties to negotiate for a reasonable settlement failed. This was not only because of the complexity of the problem, but also because Indonesia had frequently supported the Philippine government, emphasizing the danger of intervening in the internal affairs of other states. In his speech at the Sixth ICFM in Jeddah in July 1975, the Indonesian Foreign Minister, Adam Malik, asserted, 'To insist on a prior public declaration agreeing to the creation of an autonomous region, with a separate government and army, as a condition for the success of those talks, we believe, cannot be accepted by any sovereign government worthy of its name' (quoted in Philippines, Department of Public Information, 1976: 29). In recent negotiations between the Aquino government and the MNLF, Indonesia was again involved in behind-the-scenes manoeuvring (*Far Eastern Economic Review*, 11 September 1986). It was motivated mainly by the desire to preserve regional stability.

Apart from the involvement of individual Muslim countries, the Organization of the Islamic Conference has played a dominant role in the efforts to resolve the Muslim problem in the Philippines. One of the aims of the OIC, as set out in its charter of 1972, is 'to strengthen the struggle of all Muslim people with a view to safeguarding their dignity, independence and national rights' (Europa Publications, 1985: 229). The OIC, founded in May 1971, consists of the Conference of Heads of State as its supreme body and the Islamic Conference of Foreign Ministers. The ICFM meets annually to consider the means for implementing the general policy of the Organization. Table 5.1 shows the current members of the OIC.

The struggle of the Moro people attracted the attention of the OIC and in February 1972 the Third Islamic Conference of Foreign Min-

TABLE 5.1
Members of the OIC, 1985

Muslim States			
Afghanistan[1]	Gabon	Malaysia	Senegal
Algeria	The Gambia	Maldives	Sierra Leone
Bahrain	Guinea	Mali	Somalia
Bangladesh	Guinea-Bissau	Mauritania	Sudan
Benin	Indonesia	Morocco	Syria
Brunei	Iran	Niger	Tunisia
Burkina Faso	Iraq	Oman	Turkey
Cameroon	Jordan	Pakistan	Uganda
Chad	Kuwait	PLO	United Arab Emirates
The Comoros	Lebanon	Qatar	Yemen Arab Republic
Djibanti	Libya	Saudi Arabia	Yemen, People's
Egypt[2]			Democratic Republic

Source: Europa Publications (1985: 228).
Note: Observer status has been granted to Nigeria and to the 'Turkish Republic of Cyprus' which declared independence in November 1983.
[1]Afghanistan's membership was suspended in January 1980.
[2]Egypt's membership was suspended in May 1979 and restored in March 1984.

isters, in Jeddah, reviewed the situation of the Muslims in the Philippines and expressed serious concern. In 1973 in Benghazi, the ICFM passed a resolution calling for the investigation of the 'plight of Muslims living in the Philippines' and formed a ministerial committee of four, namely Libya, Saudi Arabia, Senegal, and Somalia, to visit the Philippines and look into the Muslim situation. It also established a voluntary fund to help Muslims in the Philippines, to be financed by Muslim governments as they saw fit. The ICFM also requested the governments of Indonesia and Malaysia in 1973 to exert their good offices within the framework of ASEAN for the same purpose.

In August 1973, the delegates of Foreign Ministers of the four nations visited Mindanao and Sulu. They were followed by a group of diplomats from predominantly Muslim countries and a special mission from Libya and Egypt. The Philippine government reported that the delegates were convinced that no persecution or genocide was being committed on the Muslims (Philippines, Department of Public Information, 1976: 28). Nevertheless, the ICFM meeting held in Kuala Lumpur in June 1974 adopted Resolution No. 18 (Appendix 9), calling upon the Philippine government to cease all measures that resulted in the killing of Muslims and the destruction of their property and places of worship. It urged the Marcos regime 'to find a political and peaceful solution through negotiation with Muslim leaders, particularly with the representatives of the Moro National Liberation Front in order to arrive at a just solution to the plight of the Filipino Muslims within the framework of the national sovereignty and territorial integrity of the Philippines' (ICFM, 1974). The ICFM appealed to peace-loving states, while recognizing the Moro problem as an internal problem of the Philippines, to ensure the safety of the Muslims and the preservation of their liberties. In addition, the Conference decided to create an agency called Filipino Muslim Welfare and Relief Agency to aid the Moros so as to ameliorate their plight and improve their socio-economic well-being. This was to be financed and controlled by the Islamic Solidarity Fund (ICFM, 1974).

Hence, the Kuala Lumpur conference gave both the Philippine government and the MNLF leaders a clear sign of the purpose and extent of the OIC involvement. While the MNLF leaders consolidated their strength and extended the scope of their military operations against the Armed Forces of the Philippines, the OIC continued to work for a solution to the conflict (Noble, 1984: 8). At the invitation of the Philippine government, the Secretary-General of the OIC, Dr Mohammad Hassan Al-Tohamy, visited the Philippines to discuss matters in connection with the Resolution No. 18 adopted in Kuala Lumpur. He succeeded in persuading both sides to meet in Jeddah in January 1975. The MNLF agreed to give up independence as a goal, but it demanded an internally sovereign Bangsa Moro State that would have its own security force for maintaining internal order. The Philippine panel, which comprised Executive Secretary Alejandro Melchor, Admiral Romulo Espaldon, Ambassadors Lininding Pangandaman and Pacifico Castro, Chancellor Ruben Cuyugan, Dean Cesar Majul, Colonel Jose Almonte, and economist Gary Makasiar, met the MNLF panel which consisted of Nur Misuari, Salamat Hashim, Abdul Baki Abubakar, Hamid Lukman, and Abdurasad Asani. The Jeddah meetings were conducted from 18 to 29 January 1975 in the presence of the OIC Secretary-General. The talks failed, however. The Philippine government did not agree with the MNLF leaders, who insisted on the creation of an autonomous region as a pre-condition for negotiations. In rejecting the MNLF demand, the head of the Philippine Panel, Alejandro Melchor, informed the OIC Secretary-General:

In categorical terms, what Mr. Nur Misuari and his group vaguely have in mind is not basically oriented toward the welfare of the Islamic Communities in the Philippines; it is, instead, designed to establish a new power structure and system which did not, and does not, exist, and which they hope to establish by fiat of an agreement.... Actually, Mr. Misuari and his associates have no grounds to stand on as far as their claim to having created what they like to call a 'Bangsa Moro State' is concerned (Philippines, Ministry of Foreign Affairs, 1980a: 13–14).

The Jeddah talks, which were scheduled to commence on 7 April 1975, did not occur. In the same month, President Marcos appointed a panel of Muslim leaders (Ambassador Lininding Pangandaman, Commissioner Simeon Datumanong, Brigadier-General Mamarita Lao, and Sheikh Abdul Hamid Camlian) to conduct dialogues with rebel leaders in the field. Meetings with some 500 rebels were held in Zamboanga City on 17 April and 30 June 1975 (Philippines, Ministry of Foreign Affairs, 1980a: 9). At the same time, the OIC Ministerial Committee of Four approved a nine-point agenda (Appendix 10) to be used as a basis for the resumption of negotiations. This draft agenda was accepted by the

MNLF, but was rejected by some 215 Muslim leaders, government officials, and rebels whom Marcos assembled in Zamboanga City for the second time. They argued that 'the agenda approved by the committee of four violates the sanction of the Constitution, impugns national sovereignty, territorial integrity, and insults national pride' (Philippines, Department of Public Information, 1976: 58). Instead, the group endorsed a Marcos plan to create four 'autonomous' regions in the southern Philippines that would be headed by appointees responsible directly to the President. Nevertheless, the ICFM which convened in Jeddah on 10 July 1975 approved the draft of the Committee of Four and recommended that negotiations between the two parties be arranged as soon as possible.

The newly installed Secretary-General of the OIC, Dr Ahmadu Karim Gaye of Senegal, met President Marcos at the United Nations Commission on Trade and Development (UNCTAD) in Nairobi on 6 May 1976. He informed President Marcos that 'the members of the committee felt that the outright rejection by local Muslim leaders of the Draft Agreement ... was ill-advised' (Philippines, Ministry of Foreign Affairs, 1980a: 12). A week later, the ICFM conference in Istanbul passed a resolution which called for the resumption of negotiations between the Philippine government and the MNLF based on Resolution No. 10 (Appendix 11) adopted by the Jeddah conference in 1975. In August 1976, Secretary-General Gaye of the OIC and other high-ranking officials from Senegal, Libya, and Somalia met President Marcos, who agreed to resume negotiations with the MNLF on the following conditions:

(1) It would be part of the continuing peaceful efforts of the Philippines government in negotiating with MNLF members as Filipino nationals; (2) that the meeting would not confer belligerency status on the MNLF; (3) that the national sovereignty and territorial integrity of the Republic of the Philippines are non-negotiable; and, (4) that the Islamic Conference and the MNLF should not impose any prior conditions as advanced by Misuari during the 1975 Jeddah conversations (Philippines, Ministry of Foreign Affairs, 1980a: 14).

In October, US$1 million was donated by the Islamic Solidarity Fund to the Agency for Development and Welfare of the Muslims in the Philippines (Philippines, Ministry of Foreign Affairs, 1980b: 179).

Meanwhile, the Philippine government began to strengthen its diplomatic relations with West Asia and Muslim countries. The campaign to maintain ties with Muslim states was many-faceted: it included the 'opening of formal relations with a number of Islamic countries, exchange of special high-level missions, intensified information programmes, and the crystallization of a completely new policy on the entire Middle East issue' (Philippines, Department of Public Information, 1976: 31). The realignment of foreign policy was expressed following the outbreak of the Arab–Israel conflict in 1973, when the Philippines condemned Israel's occupation of Arab territories as an act of aggression and called for the withdrawal of its forces from all occupied Arab lands in accordance with the resolution of the Security Council adopted in 1967. The Philippine government then signed a common declaration with other ASEAN countries deploring Israel's 'territorial expansion by force'. The Philippines has also recognized the PLO as the legitimate representative of the Palestinian people. In November 1973, Philippine Foreign Secretary Carlos Romulo conveyed a message from President Marcos to King Faisal of Saudi Arabia assuring him of the Philippines' continued support of the Arab cause in the Middle East conflict. Meanwhile diplomatic relations between the two countries were formalized by the appointment of Lininding Pangandaman, a Maranao Muslim, as the Philippines' first Ambassador to Saudi Arabia. Subsequently, diplomatic ties were set up with the United Arab Emirates and embassies were opened in Iran and Algeria, with concurrent accreditation in Lebanon, Kuwait, and Abu Dhabi. Efforts were made to strengthen existing relations with Afghanistan, Bangladesh, the Maldives, Pakistan, Turkey, Iraq, Jordan, Morocco, Tunisia, Nigeria, Senegal, Sierra Leone, and Somalia (Philippines, Department of Public Information, 1976: 31–2).

At the height of the separatist conflict in 1975, the First Lady, Mrs Imelda Marcos, had played an important role in strengthening Philippine relations with the Arab world. Acting as special representative of President Marcos, the First Lady's Middle East diplomacy took her first to Egypt where she had talks with President Sadat and laid groundwork for social and cultural exchanges as well as strengthening the understanding between the leaderships of the two countries. In Algeria, Mrs Marcos met President Houari Boumedienne, who expressed the hope that the Muslim problem in Mindanao would be solved through peaceful means. On the suggestion of President Boumedienne, Imelda Marcos proceeded to New York to seek the help of Algerian Foreign Minister Abdul Aziz Bouteflika (then President of the General Assembly of the UN) in presenting the Philippines' case before Arab delegates to the UN. From New York, the First Lady went to Riyadh to convey the condolences of President Marcos to the government of Saudi Arabia when King Faisal was assassinated. Such high-level diplomatic contacts opened the way for substantive demonstrations of goodwill from Arab leaders, including visits to the Philippines by King Hussein Ibn Talal of Jordan, Madame Jehan Sadat, wife of President Sadat of Egypt, President Omar Bongo of Gabon, and Sheikh Abdullah Al-Sheikh, Minister of Education of Saudi Arabia. In Tunisia, President Habib Bourguiba told Information Secretary Francisco Tatad of the Philippines that Tunisia would not support any move to dismember the Philippines.

Within that framework, Mrs Marcos visited Libya. The meeting between her and President Qadhafi resulted in a series of talks in Tripoli from 15 to 23 December 1976 between the Philippine government panel and the MNLF. The government panel consisted of Under-Secretary Carmelo Barbero, Lininding Pangandaman, Simeon Datumanong, Karim Sidri, Pacifico Castro, and Colonel Eduardo Ermita, while the

MNLF panel comprised Nur Misuari, Salamat Hashim, Abdul Baki Abubakar, and Abdurasad Asani. Libyan Foreign Minister Ali Treki presided over the talks which were conducted in the presence of the Committee of Four, including the Secretary-General of the OIC. The negotiations culminated in the signing of the Tripoli Agreement, which provided for a cease-fire and tentative terms for a settlement (see Appendix 1).

The cease-fire, which was to be supervised by a committee representing the Philippine government, the MNLF, and the Quadripartite Ministerial Commission, was signed between the two parties on 20 January 1977 in Zamboanga City. Although fighting continued in some areas, the cease-fire was generally successful until it collapsed in late 1977. To finalize the terms of the Tripoli Agreement, the government panel and the MNLF met again in Libya from 9 February to 3 March 1977. This time the negotiations were stalemated because of differences over the degree of autonomy and the role of the MNLF in its administration. Mrs Marcos was again sent to Tripoli for talks with President Qadhafi in an effort to break the deadlock. The talks resulted in the Qadhafi–Marcos understanding which covered the following points: '(1) A decree proclaiming autonomy in thirteen provinces be issued by President Marcos; (2) a provisional government be set-up; and (3) this provisional government to hold a referendum in the area of autonomy concerning the administration of the government' (Qadhafi, 1977). As a consequence, President Marcos issued Presidential Proclamation No. 1628 declaring autonomy in the thirteen provinces. Nur Misuari was offered the Chairmanship of the newly organized provisional government, but he refused because the substance of autonomy proposed was unacceptable. Marcos then appointed six Muslim and seven Christian governors and Commissioner Simeon Datumanong as the provisional government.

After having been postponed three times, the referendum was held on 17 April 1977. The Muslims generally boycotted it because Marcos proceeded to hold a referendum on critical provisions of the agreement itself (Noble, 1984: 9). The military forces were placed on combat readiness while the MNLF insisted on the creation of an autonomous region under their control (Abbas, 1979: 116–17). On 22 April, a 17-member delegation of the OIC, including Secretary-General Gaye and Libyan Foreign Minister Ali Treki, and the MNLF leaders resumed negotiations with the Philippine government in Manila in the hope of resolving their differences. However, each side refused to accept the other's compromise proposals and the talks collapsed. A communiqué issued by the OIC delegation declared that the negotiations failed because of the negative attitude of the Philippine government and the violations of the previous agreements (Abbas, 1979: 118). It warned that resumption of hostilities seemed inevitable. The Philippine government attributed the failure to reach a settlement to the unacceptable demands and the intransigent stand adopted by the MNLF. It sent Secretary of Foreign Affairs Carlos Romulo to clarify the Philippine stance in several member states of the OIC. Nevertheless, Resolution No. 7/8-P adopted by the ICFM in May 1977 considered the Philippine government responsible for the failure of the negotiations both in Tripoli in February and in Manila in April 1977. The Resolution requested the Muslim states to support the MNLF 'by all ways and means for achieving all demands of Muslims in South Philippines' (ICFM, 1977a).

Thereafter, the OIC continued to denounce the Philippine government for avoiding its international obligations. As the cease-fire ended in late 1977, the ICFM meeting in Dakar in 1978 called on both parties to end all fighting, respect the cease-fire agreement, and resume negotiations. However, the Marcos regime no longer paid much attention to the call of the OIC; its bargaining position had been strengthened by favourable international circumstances and a marked decline in the strength of the MNLF.

To keep the Moro struggle alive, the OIC decided in 1977 to give the MNLF observer status within the Conference (ICFM, 1977b). From then on the annual meetings of the OIC have essentially reiterated the Organization's calls for negotiations to implement the Tripoli Agreement. At the same time, the Islamic Solidarity Fund, a subsidiary organ of the OIC, continued to provide financial support to the MNLF in order to keep the Moro struggle going. Thus, when the Aquino government expressed its willingness to meet with the MNLF leaders, the OIC was more than willing to assist in ending the 14-year old hostilities. The Philippine's neighbours, Indonesia, Malaysia, and, to some extent, Singapore, also used their influence in persuading the two parties to come to the new negotiating table (*Far Eastern Economic Review*, 11 September 1986).

In the final analysis, external involvement by Muslim states and the OIC had a significant effect on the Moro separatist struggle. Notwithstanding the ideological differences which have led Islamic states such as Libya, Saudi Arabia, and Kuwait to direct their support to different factions of the movement, the support of Muslim countries worldwide has sharpened self-confidence and increased the level of fighting that culminated in the Tripoli Agreement on terms much more favourable to the Moros than their leadership ever would have gained without outside assistance. Furthermore, external help had kept the movement alive, at least at its minimum capacity. On the other hand, outside factors shaped the goals and tactics of both conflicting parties and persuaded them to enter peace negotiations.

Government Responses

In response to the separatist threat posed by the Muslims in the south, the central government in Manila employed both military and nonmilitary measures. Like other governments, the Marcos administration believed that such a threat could not be permanently quelled without addressing successfully some of the socio-economic and political dimensions from which it sprang.

In general, the Moros in Mindanao were seen by the Philippine government as a problem because they were 'backward' and 'unwilling' to accept changes. The problem, however, was regarded as manageable. A multifaceted policy was devised in an attempt to promote Moro integration, primarily through socio-economic development programmes, and to maintain peace and order in the region through the use of coercive measures. But the latter was pursued with far more vigour than the former.

The use of coercive measures by the government began when the Philippine Constabulary (PC) started to support the 'Ilaga terrorist squads' fighting against the Moros in 1968. By the middle of the 1970, the PC became very active on the side of the Ilagas; in September 1971 many Moro villagers on the island of Mindanao were subjected to the PC's 'search and destroy' missions; and by March 1972, 12 out of 35 municipalities in Cotabato province alone were under PC control, causing thousands of Muslim villagers to flee the areas (Ahmad, 1980: 32; *Manila Times*, 21 March 1972). After the declaration of martial law in 1972, full-scale military operations to suppress the Moro resistance movement were initiated. The main objectives of the campaigns were to conduct 'search and destroy operations against defiant rebel groups, to impose the government's will in the affected areas ... and in the process assist in the restoration of local governments' (Philippines, Department of National Defense, n.d.: 4). Among the more important military operations documented by the Department of National Defense were 'Operation Sibalo', 'Operation Reina Regente', 'Operation Pamukpok', 'Operation Lebak', 'Operation Batikus', and 'Operation Bagsik'.

Operation Sibalo on the island of Jolo was a campaign to control Sibalo Hill, which was considered tactically vital to military operations as it commands the surrounding areas. The control of the hill would furthermore divide the rebels into two separate groups. The forces participating in this operation comprised two marine companies, one infantry company, two mortar platoons, and two Air Force helicopter units, totaling about 500 military men. The campaign lasted two days, from 28 to 30 December 1972. No significant resistance was made by Moro separatists. The government viewed Operation Sibalo as a major psychological gain for the AFP troops, because it revived the offensive spirit and demonstrated to the populace the serious intention of the government to assert its authority (Philippines, Department of National Defense, n.d.: 95–109).

Operation Reina Regente was launched by the Central Mindanao Command (CEMCOM) under Brigadier-General Fortuna Abat. Its aim was to 'flush out all the insurgents from their bases of operation' (Philippines, Department of National Defense, n.d.: 84), especially at Reina Regente Mountains, Datu Piang, Cotabato. To ensure the success of this campaign, Brigadier-General Abat employed a force of more than 2,000 men, comprising the 6th Infantry Brigade, Army Artillery Group, Combat Air Strike Force, a PC company, and all Civilian Home Defense Forces in the area. The operation began on 28 January and ended on 5 April 1974. The main effort of the operation was to decimate some 800 armed separatists under several leaders, including Abdulatip Mohammad, Santiago Kadatuang, Ustaz Daudayuan, Subo Dalamdas, Duskan Talipasan, and Kinok Dundling (Philippines, Department of National Defense, n.d.: 75–94). The campaign was reported by the CEMCOM to have been successful.

Operation Pamukpok was a campaign which attempted to capture Barrio Tuburan in the Lamitan district, Basilan, a stronghold of an undetermined number of Moro rebels. Several operations by government ground troops had been launched but Tuburan remained under control of the rebels. In view of the threat posed by the separatists, an amphibious operation was decided upon. The main forces that participated in the campaign were Amphibious Task Force 32, 2nd Battalion Landing Team, Naval Attack Force 32, Mortar Platoon, and Naval Gunfire Support Group. The ten-day operation, from 5 to 14 July 1973, accounted for about 200 rebels killed and many casualties and homeless. The capture of Tuburan was considered highly successful, preventing Muslim separatists from turning it into an almost impregnable bastion where they could offer strong opposition (Philippines, Department of National Defense, n.d.: 57–74).

Operation Lebak was a campaign to destroy insurgent forces in Tran and Lebak, Cotabato. The group was under the command of Datu Sangki Koran, former Councilman of Lebak, and Pendi Koran Omar, Koran's deputy. The operation lasted six months, from March to August 1973. It comprised Task Forces 'Cosmos' and 'Sarsi' which were composed of the 1st Composite Infantry Battalion, the 21st and 22nd Infantry Battalions, three Constabulary companies, and the Civilian Home Defense Forces. The six-month battles between the AFP and the separatists cost hundreds of life and thousands became homeless (Philippines, Department of National Defense, n.d.: 45–56).

Operation Batikus was a campaign at Siasi Island launched in an attempt to prevent separatist guerrillas capturing Siasi town, the only remaining government strong position in the island at the time. The operation was composed of Naval Task Force 32, Marine Landing Force 33, and Tactical Air Support units. A three-day operation, from 14 to 16 August 1973, destroyed many villagers' houses and boats; the separatists suffered at least 85 killed in action and an undetermined number were wounded. The government casualties were counted at only 2 killed and 6 wounded (Philippines, Department of National Defense, n.d.: 19–44).

Operation Bagsik was an operation to recapture the city of Jolo from the MNLF. The city of Jolo was under the control of Muslim separatists for three days before Operation Bagsik was launched. More than 2,000 AFP troops, including air and naval support, attacked the city from 4 February to 10 April 1974. The mission was accomplished, but the toll was heavy. The government recorded 99 AFP troops killed in action and 22 wounded, while separatist casualties stood at 516 killed and 224 wounded (Philippines, Department of National Defense, n.d.: 1–18). The city centre was almost completely destroyed.

These operations give an idea of the nature and extent of the armed conflict, which had developed into a conventional war between the Marcos regime and Moro separatists under the leadership of the MNLF. The war reached its peak and became stalemated in 1975. As discussed in Chapter 3, the costs of the war were considerable and attracted sympathetic attention from the international community. In 1976 the two parties signed the Tripoli Agreement.

With respect to socio-economic development, several steps had been taken from as early as the 1950s. A Commission on National Integration (CNI) was created in 1957 which was designed to foster 'the moral, material and political advancement' of the non-Christian Filipinos. The Mindanao State University (MSU) at Marawi City was established in 1955 to promote education among the peoples of the south, particularly the Moros. And in 1961 the Mindanao Development Authority (MDA) was formed to accelerate the development of the region. However, these programmes fall far short of their objectives due to the complexity of the problem and because of financial and management difficulties. The notable accomplishment was the CNI's scholarship programmes for members of the minorities. Between 1958 and 1967, for example, about 8,300 scholarships were awarded (see Table 4.1). Following the reports of a Senate committee on national minorities in 1963 and 1971, which identified resettlement and land-grabbing as the major sources of conflict in Mindanao, the Marcos government announced a number of social and economic programmes as part of a package intended to win over dissident Moros (May, 1985: 113–14; Mastura, 1980).

As Moro–Christian relations deteriorated in the early 1970s, the armed forces were sent in to quell the situation. But the military became a catalyst for further violence because of their bias against Muslims and their indiscipline. Along with coercive measures, President Marcos had also issued a series of decrees, orders, proclamations, and letters of instruction aimed at ending the conflict. Some of the early decrees were intended to remove the restrictions on the traditional barter trade (Presidential Decree 93, 1973); to grant amnesty to persons who had committed any act penalized by the existing laws in Muslim areas (PD 95, 1973); to authorize the use of Arabic language as a medium of instruction in *madaris* (Letter of Instruction 71-A, 1973); to declare Muslim holidays as legal Philippine holidays (Proclamation 1198, 1973); to direct the University of the Philippines Board of Regents to establish an Institute of Islamic Studies (LOI 82, 1973); and to declare ancestral lands occupied and cultivated by national cultural minorities as alienable and disposable in 26 provinces throughout the country (PD 410, 1974). More significantly, the government claimed to have allocated about 818 million pesos for development programmes aimed at creating conditions for closer national integration and at securing lasting peace in the area (PTF-RDM, 1973: 5–6).

Towards this end, the Rehabilitation and Development programme (RAD) for Mindanao was formed. It covered the provinces in which the majority of Muslims reside. The RAD comprised three broad types of activities: rehabilitation, reconstruction, and development. The first two included those activities required to restore normalcy in the region, such as assistance to evacuees and repair of damaged infrastructure, while the third involved such activities as expansion of agricultural and industrial production and improvement of education and health services (Melchor, 1973: 66).

To implement this programme, the Marcos regime created the Presidential Task Force for the Reconstruction and Development of Mindanao (PTF-RDM) headed by Executive Secretary Alejandro Melchor. The objectives of the PTF-RDM, as defined by Executive Order No. 411 of 1973, were 'the assessment of damage on private property, mobilization of funds and preparation of an integrated programme of full reconstruction, and restoration of peace and order' (PTF-RDM, 1973: 5).

The first of the RAD's rehabilitation programmes (LOI 30, 1972) co-ordinated through the Special Programme of Assistance for the Rehabilitation of Evacuees (SPARE) was to assist displaced families from areas of disturbance. The affected villagers were to be resettled in their homes as soon as peace and order were restored. The projects pursued under LOI 30 of 1972 were for housing, extension of agricultural credit, and provision of technical assistance to farmers and evacuees.

With respect to reconstruction and development, the main focus was on reconstructing damaged facilities such as roads, bridges, and ports and restoring production to pre-crisis levels. Efforts were made to increase vegetable, rice, and maize production to meet the needs of the region. Small-scale industries were encouraged through financial assistance from such institutions as the Development Bank of the Philippines and the Private Development Corporation of the Philippines. Meanwhile, the education system was being re-oriented to develop skills to enable people to exploit the natural resources of the region; attempts were also made to improve health services through rural health and disease control programmes.

These RAD programmes exemplified the government's conception of the problems and its attempts to solve them along socio-economic–cultural lines. Unfortunately, while the policy was well conceived, its implementation was not sufficient to redress the social and economic imbalances resulting from long neglect and discrimination against the Muslims. The PTF-RDM was later incorporated into the Southern Philippines Development Administration (SPDA) together with such other programmes as the SPARE, CNI, and MDA.

At the height of the armed confrontation, when at least one-third of the entire Armed Forces of the Philippines was operating in Moroland, the national government set up the Southern Philippines Development Administration (PD 690, 1975). The SPDA was charged with 'the main responsibility of promoting economic development and social stability through corporation and non-corporation ventures' (Philippines, Department of Public Information, 1976: 8). The corporate projects were essentially profit-oriented, aimed at generating funds that would directly or indirectly benefit the population through job opportunities and taxes for local governments in the region. The non-corporate programmes were public-service-oriented, designed to enhance social development.

The administration of the SPDA, which was headed by a board of directors appointed by President Marcos, was involved in a four-pronged development effort: resource development, social-oriented activities, financial investments, and technical assistance. Resource development involved the use of the region's natural resources for profit-oriented projects. The social-oriented activities included projects in education, land reform, human settlements, and health which were implemented jointly by various government agencies. The last two became operable when the SPDA entered into joint ventures with public or private institutions (Philippines, Department of Public Information, 1976).

In addition, the SPDA was entrusted with the responsibility of providing help and rehabilitation facilities for Muslim ex-rebels and evacuees in Mindanao. In 1980, for instance, a special fund of 25 million pesos was to be administered by the SPDA for a rehabilitation programme for the MNLF and other groups working with the government development effort (May, 1985: 114).

Other major programmes designed to promote and accelerate the socio-economic growth and development of the Muslims in Mindanao included the creation of the Philippine Amanah Bank (PAB), the codification of Muslim laws, and the establishment of the Ministry of Muslim Affairs (MMA). The PAB was founded to meet the banking, credit, and financial requirements of Muslims. As a specialized bank, it operated on a dualistic basis: conventional and Islamic. With respect to the latter, the PAB developed the Islamic concept of banking which is based on no-interest and partnership principles. In Islam, interest (*riba*) is not permissible; it should be waived in favour of profit-sharing. As a consequence, the PAB included in its organizational structure a Muslim Development Fund (MDF) to which the waived savings interest would be paid and in turn be used for various projects geared towards developing Muslim areas and promoting the welfare of the Muslims (PTF-RDM, 1973: 48–51). Another direct service of the bank to the Muslim community was the Pilgrim's Special Saving Deposit (PSSD). The PSSD was designed to help potential Muslim pilgrims to save for the *haj* and to give them an opportunity to increase their savings through PAB's Islamic investment scheme (Mastura, 1984: 266).

The codification of Muslim Personal Laws (PD 1083, 1977) was an outcome of the integration strategies that the Marcos administration had pursued in an attempt to incorporate some aspects of Islamic law (e.g. marriage, divorce, personal status, property relations, will succession and inheritance, gifts, wage, and endowments) into the civil laws of the Philippines. As President Marcos stated, 'The Muslim heritage is part of the heritage of the nation. Their laws should be part of the law of the land' (PTF-RDM, 1973: 32–3). Thus, Presidential Decree No. 1618 was promulgated in 1979 to establish, *inter alia*, Sharia Courts in the Autonomous Governments of Regions IX and XII (see below). Four years later, the Philippine Supreme Court issued rules of procedure for these courts and for a Bar examination to provide them with qualified personnel. The courts were expressly made applicable only to Muslims (see Philippines, Ministry of Muslim Affairs, n.d.; Mastura, 1984: 199–212). Unfortunately, the courts have not yet become organized and operative as of 1989.

In 1981 the special status of the Muslims was recognized by the national government with the creation of a Ministry of Muslim Affairs (Executive Order 697) 'to insure the integration of Muslim Filipinos into the mainstream Filipino society with due regard to their beliefs, customs, traditions and institutions' (Wurfel, 1985: 225). However, the appointment of Admiral Romulo Espaldon, who was a recent convert to Islam, as head of this new ministry was not acceptable to many Muslims, who did not regard Espaldon as their leader. This made the ministry less effective as an integrating mechanism than it might have been. In 1985 Espaldon was replaced by a Maguindanao traditional leader and politician, Simeon Datumanong. After the Aquino government came to power in 1986, the ministry was headed by another Muslim politician, Condu Muarip from Basilan. It has been suggested that the creation of such agencies by the Marcos regime was aimed not so much at socio-economic reforms beneficial to the Moro masses, but at offering the traditional Moro leaders 'a share in the spoils of martial law' and so inducing some members of the traditional élite and their followers who had joined the separatist movement to surrender (Nemenzo, 1985: 240–6).

In addition to these projects, 'President Marcos declared all unappropriated agricultural lands of the public domain being occupied and cultivated by Muslims and other national cultural minorities as ancestral lands and therefore alienable and disposable' (Philippines, Department of Public Information, 1976: 20). These ancestral lands, located in 26 provinces (including 11 Muslim provinces), were subdivided into farm lots not exceeding 5 hectares and allocated to members of the national cultural communities who had been cultivating them at the time of the issuance of Presidential Decree (PD 410, 1974).

This land reform was intended to appease Muslims and other cultural minorities who resented the detachment of their lands from the traditional pattern of community and clan ownership. However, the weakness of the land decree was that its implementing order created bureaucratic procedures which ironically made it more difficult for cultural minorities to secure land titles. As Lynch has suggested, efforts to preserve ancestral land have proved inadequate, and the land continues to be usurped at an increasing rate (quoted in Wurfel, 1985: 226; see Dumarpa, 1983). Government projects and large corporations have managed to gain access to public land, regardless of ancestral occupancy. For instance, dam projects in Mindanao are expected to consume about 126 000 hectares of land and displace nearly half a million people (Silva, 1979; Mercado, 1981; Wurfel, 1985: 226). Furthermore, Presidential

Decree 1559 issued in 1979 directed that squatters, cultural minorities, and other occupants of forest and unclassified public land shall be ejected and relocated if the land is to be used for other purposes as determined by the Bureau of Forests.

Another major response of the Marcos administration was the creation of local autonomy based on the Tripoli Agreement. The establishment of the Autonomous Governments of Regions IX and XII was an effort by the central government to find a solution to the separatist conflict within the context of national sovereignty and the territorial integrity of the Republic of the Philippines. Under the Tripoli Agreement certain rights and prerogatives were granted to the thirteen Muslim-populated provinces in the south. Among them were the rights to have their own administrative, financial, and economic systems in accordance with the objectives of the autonomy, to set up their own Sharia Courts, and to have Special Security Forces in the region (see Appendix 1). A Legislative Assembly and an Executive Council were to be formed in the autonomous regions.

However, President Marcos implemented the agreement on his own terms. In May 1977, Presidential Proclamation 1628-A was issued, proclaiming the adoption of the outcome of a controversial referendum–plebiscite of April 1977. In July 1979, the Legislative Assembly (Sangguniang Pampook) and Executive Council (Lupong Tagapagpaganap Ng Pook) were created by virtue of Presidential Decree 1618. The decree was widely protested by the MNLF because it departed from the Tripoli Agreement and did not grant the autonomy which the MNLF claimed had been agreed. For example, two autonomous governments had been established, instead of one, and the exclusion of the provinces of Davao del Sur, Palawan, and South Cotabato from the autonomous regions violated the accord. Moreover, the legislative powers granted to the autonomous governments were negated by Section 4 of PD 1618 which prevented them from acting on matters that were within the jurisdiction and competence of the national government; Section 35 of the decree stated that 'with respect to legislation, national laws shall be supreme vis-a-vis regional laws enacted by the Sangguniang Pampook'. The decree empowered regional assemblies to impose taxes and fees, but the Ministry of Finance guide-lines limited the revenue-raising powers of the regional governments to inconsequential matters, such as signboards and secretary's fees, which if imposed would not have accumulated tax revenues of more than 100,000 pesos annually. This was not the financial system laid down by the Tripoli Agreement. Not surprisingly, Nur Misuari boycotted the Regional Assembly election, despite the invitation by President Marcos to lead the restructured regional autonomous government. Salamat Hashim also refused to accept the government offer to take Misuari's place (May, 1985: 116). In 1982 Abdul Khayr Alonto (former Vice-Chairman of the MNLF), who became the Speaker of the Region XII Assembly, published a booklet, addressed to President Marcos, which criticized the operation of the autonomous governments and called for the merger of the two regions and the granting of 'a meaningful autonomy'. Alonto was dropped from the Marcos KBL ticket for the 1982 Regional Assembly election (May, 1985).

The Autonomous Governments of Regions IX and XII, in fact, did not have any real legislative powers; they could 'only pass resolutions addressed to the President or the respective heads of ministries, bureaux, and other national offices requesting action or appealing for aid without any legal sanction to enforce its measures' (Pangarungan, n.d.: 6). Some Muslims observed with sarcasm that a town or even a *barangay* council possessed more power than the Regional Assemblies in that the former could enact ordinances with the force of law carrying penalties in case of violation. Many concluded that the terms of the Tripoli Agreement had not been fully implemented.

Despite a claim by President Marcos that his policies had 'effectively terminated' the MNLF and that more than 26,000 rebels had returned to the folds of the law, the separatist struggle persisted. The use of military forces to suppress the movement, as described above, remained the preferred method.

The failure of the Marcos administration to resolve the Moro problem through non-military measures can be attributed to a number of factors. It has been alleged that while, nominally, government programmes were mostly aimed at socio-economic reforms that would benefit the Moro masses, in practice they were promoting the interests of those Moro leaders who were pro-government. Commenting on the government's socio-economic development programmes, Muslimin Sema, a Maguindanao rebel leader, said that 'The government may build golden bridges to span the rivers of Mindanao and Sulu, cement roads to criss-cross their forests and plains, but what good are these roads and bridges if the Muslim people cannot use such roads and bridges freely because of their being Muslims' (quoted in Glang, 1973: 39).

Secondly, most government policies with respect to minority communities are perceived by the Muslims as instruments of assimilation. Assimilation into a 'monolithic Filipino nationhood' is what the Moros have been trying to avoid for centuries, fearing the loss of their identity as the Muslim *ummah*.

In spite of all efforts to advance the socio-economic and political status of the Moros (no less than 60 Presidential Decrees, Executive Orders, and Proclamations had been declared from 1972 to 1982), the separatist conflict remains. Since the Aquino government replaced the Marcos regime, negotiations with the MNLF have been initiated in an attempt to end the conflict. Moreover, the Constitution of the Republic of the Philippines, which was ratified by referendum in February 1987, has adopted general provisions for the creation of autonomous regions in Muslim Mindanao and in the Cordilleras. However, what kind of autonomy is intended by the new constitution remains to be seen. Section 17 of the autonomous provisions states, 'All powers, functions, and responsibilities not granted by this Constitution or by law to the autonomous regions shall be vested in the National Government' (*Constitution of the Philippines 1986*, Article X).

Be that as it may, one of the major causes of the separatist movement has been the Moros' desire to preserve their identity based on Islam. Over a period of time, Moro identification has evolved to a point where rectification of the socio-economic and political ills is no longer sufficient as a basis for accommodation. This is because 'the Islamic identity itself has taken on a life of its own' (Ayoob, 1984: 266–8).

Concluding Remarks

The involvement of Muslim states and organizations in the Moro liberation struggle helped to strengthen the separatist movement's military and political capabilities to the extent that it was able to put pressure on the central government to make some concessions and restrain its policies. The creation of the autonomous governments and other governmental agencies was part of the Manila response to such pressures. Suhrke and Noble (1977: 18), however, suggested that external involvement was 'sufficient to have a nuisance value but insufficient to compel a settlement satisfactory to one or all of the local protagonists'. Their involvement has been limited by concern over their own interests.

External involvement has been differently motivated. Malaysia's involvement, for instance, was a direct reaction to Philippine policies related to the claim to Sabah. It was supplemented by the Islamic fervour of the Sabah government under Tun Datu Mustapha and the Sabahans' ethnic link with the Moros. Libya's support was originally motivated by Qadhafi's Islamic revolutionary zeal and his anti-establishment inclination. As the consequences of the conflict became apparent, Libya changed its position from military assistance to diplomatic efforts that, through the co-operation of other states, led to the signing of the Tripoli Agreement. In the case of Muslim international organizations, such as the OIC and the Muslim World League, their involvement was motivated by a concern over the welfare of Muslim brethren, especially those who were considered to be mistreated by non-Muslims. These Muslim organizations attempted to find a settlement favourable to the Muslims. Since the demand involved separation, touching one of the most sensitive issues facing the modern nation-state, they hesitated to employ all measures possible for the settlement.

The Philippine government, on the other hand, responded to external involvement by broadening its diplomatic relations with the Muslim states to win their goodwill, while initiating various development programmes in Mindanao. However, the government responses have not been sufficient to unravel the 'Moro problem'. The roots of the problem are not only socio-economic but also ethno-cultural. Indeed, the Moro struggle has always been a struggle to defend Islam, a basic foundation of the Moro ethno-cultural identity. Such strife is regarded by the Moros as, to use Darwin's words, 'the struggle for existence'.

Thailand

External Influences

The separatist struggle of the Malay-Muslims in southern Thailand has, since its inception, involved neighbouring Malaysia. The Malays in Patani expect their kin across the border, who share a common culture, religion, and history, to come to their rescue. But their expectations have not been realized because Malaysia has had to face its own internal problems. As Suhrke and Noble (1977: 15) wrote, 'Ethnic ties—however strong—are likely to be subordinated to other considerations.' And geographical proximity paradoxically has restrained Malaysia from providing meaningful support to the Patani separatist effort.

At the beginning of the twentieth century, the royalty of the Muslim provinces solicited the British authority in Malaya to assist Patani's attempt to free itself from Thai domination (see Chapter 2). As a result, the Governor of the Straits Settlements, Sir Frank Swettenham, initiated in 1901 a move to incorporate Patani into British Malaya. However, it was aborted by London (Koch, 1977: 74–81). The Patani issue re-emerged as the British were planning post-war security arrangements for South-East Asia. Sir George Maxwell of the Colonial Office even suggested the annexation of the region. Tengku Mahmud Mahyuddin was said to have been assured by the British of their intention to incorporate Patani into British Malaya (see Omar Farouk, 1984: 243–4).

But the hopes of the Patani people were shattered (Thompson and Adloff, 1955: 161) when an Anglo-Siamese agreement for joint control of the border was signed in January 1949. With this agreement, it became clear that the people of Patani had overestimated the capacity and willingness of British Malaya to help them liberate themselves from Thailand. Even after this, Muslim leaders of Patani remained optimistic that the authorities in Malaya would eventually intervene and support their cause. Indeed, the Malay media gave wide and sympathetic coverage to the Patani plight while certain Malay political parties and personalities expressed their support (see M. Noordin Sopiee, n.d.: 10–34, 44–78). These gestures were taken seriously by the Thai authorities, who saw it necessary to invite an Alliance Party delegation to Bangkok to obtain promises of non-interference in Patani in return for funding assistance to the party (Omar Farouk, 1984: 244).

When independence was finally granted to the Federation of Malaya without any reference to Patani, many Muslim leaders in the region realized that hopes of irredentism had come to an end. Some of them began to prepare themselves for integration into the Thai nation by changing their Malay names to Thai and sending their children to Thai schools. At the same time, a large number of Patani Muslims moved to Malaya, especially to the northern states of Kedah, Kelantan, Perak, and Trengganu, while others sought refuge in Saudi Arabia. These overseas Patani Muslims later provided valuable support for the separatist fronts organized to challenge the Thai authorities.

The support of some Malay political parties for the Muslim struggle in Patani continued after independence. When the formation of Malaysia was being discussed in the early 1960s, Partai Sosialis Rakyat Malaysia (PSRM) took the view that the four Muslim provinces of southern Thailand should be included in the new proposal. Another strong supporter has been the Partai Islam se-Malaysia (PAS). PAS leaders, some of whom are familiar with the separatist movement through their association with Patani leaders from the different fronts, have often expressed support in their individual capacity to their preferred separatist organizations. Likewise, some Patani Muslim leaders who hold Malaysian citizenship joined PAS. The PAS interest in Patani was evident when the issue was raised several times in the Malayan Parliament by PAS members. At an election rally in Kelantan in 1969, Dato Mohammad Asri Haji Muda, PAS President, 'openly discussed the prospect of an alternative Malay nation—comprising the Malay states of Malaya and those of South Thailand—should Malaysia collapse as a country' (Omar Farouk, 1984: 245). In June 1974, Dato Asri, then a cabinet member in Tun Abdul Razak Hussein's administration, was accused by the Thai authorities of supporting the separatist movement and interfering in the internal affairs of Thailand when, at the PAS annual congress, he declared: 'In our considered opinion, the demand for autonomy subject to certain conditions, for the southern Thai provinces, which the liberation front has put forward, deserves to be given a favourable reception. It could constitute a sensible step towards peace and tranquility' (*Berita Harian*, 14 June 1974; Kershaw, 1985: 2). This statement brought a swift and strong response by the Thai government. Almost immediately after a Thai newspaper (*Prachatipatai*, 15 June 1974) reported the incident, the Malaysian Ambassador to Thailand was asked to clarify the situation to the Foreign Ministry. A protest to the Malaysian Embassy in Bangkok was organized and a rally was staged in the border province of Narathiwat. All sorts of allegations against Malaysia were made. For instance, *Daw Siam* (26 June 1974) accused the Malaysian government of encouraging the Muslims in the four provinces to assume dual citizenship for political purposes. On 20 June 1974, Thai legislative leaders called an emergency meeting to discuss the incident and condemned Asri as supporting the separatist aspiration of the Muslims (see Kershaw, 1985).

Apart from the political parties, the Malay villagers, especially in Kedah, Kelantan, Perlis, Perak, and Trengganu, have always been strong supporters of the Patani struggle. In 1977 a survey of Malaysian attitudes towards the Muslim problem in Patani found that 46 out of 56 respondents (82.1 per cent) would support a policy of Malaysian government intervention in favour of the Patani Muslims in South Thailand. Five (non-Malay) respondents (8.9 per cent) were opposed to any interference by the Malaysian government, while the other five (non-Malay) respondents held neither view (Omar Farouk, 1984: 255). This indicates that many Malays in Malaysia remain concerned over the fate of their ethnic brethren across the border.

Paradoxically, while many of the Malay population are sympathetic to the Muslim struggle in Patani, the Malaysian government has in many ways co-operated with the Thai government to suppress it. For example, when PAS members in the Parliament attempted to condemn the Thai government's repressive actions against Muslims in the four provinces, government authorities and parliamentarians defended Thailand, placing the blame on the communists for creating unrest in the area (Omar Farouk, 1984: 241). In an effort to relieve pressure on Thailand, the government of Malaysia in 1961 encouraged those Patani Muslim leaders who were dissatisfied with the Thai government to settle in Malaya. Similar policies have been pursued by successive Malaysian governments, resulting in thousands of Muslims migrating from Patani to Malaysia.

Malaysia has always maintained good relations with its northern neighbour. The two countries claim to uphold the principles of peaceful coexistence and mutual respect for each other's sovereignty. There has been a degree of bilateral co-operation, including limited military operations along the 600-km border. In 1977, for example, a series of co-ordinated Combined Military Operations was carried out by the armed forces of both countries: 'Operation Daoyai-Musnah' in the Sadao area in Songkhla Province and 'Operation Cahaya Bena' in the east of the Betong Salient in Yala Province. The establishment of the Thai–Malaysia General Border Committee, which functions as a forum to discuss security and related problems in the border area, has served to promote contact and communication between Thai and Malaysian officials. Official and unofficial visits by leaders from both sides have strengthened the bilateral ties and cordial relations have been further cemented by the ASEAN spirit of solidarity.

Nevertheless, the underlying suspicions that have arisen from the ethnic unrest in the Muslim provinces have not been quelled. The Thai authorities apparently are not convinced that the Malaysian government does not assist the Muslim separatist struggle. Malaysian authorities, on the other hand, believe that Thailand has not done all it can to help suppress the CPM guerrillas of different factions who use Thailand as their sanctuary. In fact, the presence of the CPM in the four border provinces of southern Thailand is allegedly considered by the Thai authorities to be politically useful as a countervailing force or bargaining chip to balance the presence of the Muslim separatist fronts in Malaysia. As Suhrke (1975: 197) observed, 'The intergovernmental solidarity reflected a simple trade-off pattern whereby support for, or acquiescence in the existence of, one rebel group was incompatible with a similar attitude toward the other rebel group.' In other words, the Malaysian government has no choice but to restrict its involvement in the Patani movement; on the other hand, it has considerable incentive to keep the conflict from being settled.

Other external factors helped to promote separatist conflict in Patani. The resurgence of Islam on the international scene, which resulted in the emergence of a Muslim bloc and the establishment of various Muslim institutions and organizations, has undoubtedly encouraged the Muslims in Patani to carry on their liberation struggle. Leaders of the different fronts have lobbied the Islamic Conference of Foreign Ministers at various meetings since the first, held in Jeddah in 1970 (BNPP, 1976). In Kuala Lumpur in 1974, the BNPP under the leadership of Tengku Abdul Jalal called on the ICFM to help secure Patani independence and to effect an oil embargo against Thailand (*Bangkok World*, 26 June 1974; Omar Farouk, 1984: 250; 1986). At the Seventh ICFM in Istanbul, the BNPP (1976) and other fronts were allowed to attend the conference as unofficial observers; the problems of Patani and of other Muslim minorities were discussed behind closed doors. They were subsequently discussed at other international Islamic forums, such as the Islamic Council of Europe in 1978. In 1980 the BNPP also submitted a memorandum on the Patani case to the Eleventh ICFM in Islamabad (BNPP, 1980). International commissions, often with the consent of the Thai government, visited the Muslim provinces for fact-finding tours (Omar Farouk, 1984: 250) and certain countries of the OIC promised to provide funds for the development of the Muslims in the south (see *Bangkok Post*, 3 August 1974; *Nation*, 23 August 1974). Contacts with the Muslim World League were made. In 1977 a letter of complaint was submitted by the BNPP to the United Nations Office in Geneva through the League (BNPP, 1977). Through these external contacts, the Patani issue began to be heard among the Muslim states, though it gained no official recognition.

As a result of these representations, various forms of assistance have been given to the fronts. Several Muslim states, including Malaysia, Saudi Arabia, Libya, and Syria provide sanctuary and/or training. Funds are sometimes given by the various governmental and private institutions and individuals to different separatist groups. These institutions include the Muslim World League in Mecca, Darul Ifta in Riyadh, the Islamic Solidarity Fund in Jeddah, Al-Auqaf in Kuwait, World Islamic Call Society in Libya, and the Baath Party in Syria. The Muslim World League has assisted the fronts several times in their drive for unity and in 1979 was able to persuade the BNPP and PULO to sign a temporary agreement to co-ordinate their operations.

Apart from direct assistance to the fronts, Muslim countries have provided scholarships and places for tertiary education to the Malay-Muslims. Many young Muslims prefer to study abroad, particularly in Saudi Arabia, Malaysia, Egypt, Libya, Pakistan, Indonesia, Kuwait, Syria, Sudan, and Iraq. This is contrary to common assumptions among the Thais that the Malay-Muslims generally lack educational motivation. Most of these overseas students are sympathetic to or are members of separatist organizations whose activities help to expose the Malay-Muslim grievances to other sympathetic Muslim organizations abroad. Even among the basically apolitical students, their perceived status as 'second-class citizens' causes them to resent the Thai authorities. In addition, the existence of a large community of Malay-Muslims in Saudi Arabia and Malaysia who continue to maintain an interest in the affairs of their homeland has been a source of finance and manpower for the separatist activities abroad.

When Malay-Muslims from the four provinces come to Mecca during the *haj* season, attempts are made by the fronts to recruit and indoctrinate them, and programmes of action are formulated during this period. Thai authorities are aware of such activities, and since the second half of the 1970s the Thai government has attached an intelligence section to the Thai Embassy in Jeddah specially to monitor the activities of the separatist fronts in Saudi Arabia and neighbouring Arab countries. The Thai government has also imposed strict clearance procedures for Muslims from the southern provinces who want to perform the *haj*.

With finance, educational facilities, and political and military training available to them, the Patani community and students abroad continue to conduct their political and diplomatic campaign against Thailand. Despite the fact that most of the Muslim states and organizations do not condone the avowed separatist objective of the Patani movement, their tacit support is motivated by the desire to preserve the *ummah* and to see that their Muslim brethren are properly treated. The OIC seminar on Muslim minorities in Perth, Australia, in September 1984, for instance, recommended that the OIC give full moral and financial support to rehabilitate the indigenous Malay-Muslim language and culture and to initiate and support economic development projects for the long-term well-being of the Patani Muslims (OIC, 1984: 7).

It is important to note that there is no official co-operation between the Moro and the Malay movements, despite the fact that they have received support from the same sources overseas. The links between them seem to be confined to formal visits and invitations to one another's offices abroad, during which they exchange ideas and information. However, no leaders of either movement ever pay official visits to each other's homeland, let alone help one another in combat. This lack of co-operation can only be explained by the fact that both movements prefer to concentrate their weak resources on obtaining assistance from the Muslim community overseas rather than on seeking to help each other.

Government Responses

The Malay-Muslim armed separatist movement in the four border provinces is regarded by the Thai central government as a threat to national integrity which should not be tolerated. The response of the government is essentially one of continuing efforts to assimilate Malay-Muslims into the Thai-Buddhist polity. This policy has been carried out more actively by the Thai bureaucratic polity that emerged following the 1932 Revolution (see Mokarapong, 1972). As discussed in the previous chapters, the 'forced assimilation policy', which was pursued during the Phibun government (1938–44), achieved adverse results and left a legacy of even deeper resentment against Bangkok (Pitsuwan, 1982: 167; Snitwongse, 1985: 263). Of particular significance was the establishment of the Sharia Offices in the four Muslim provinces in 1946 where Muslim judges were appointed to sit alongside Thai judges to deliberate on cases concerning Muslim inheritance and marriage. This was considered by many Muslims as government intrusion on Muslim legal domain because prior to the establishment of the Sharia Offices, Islamic laws in the Patani region were enforced by *ulama* without any institutional arrangement (for a detailed account on the establishment of the Sharia Offices in the four provinces, see Pitsuwan, 1982: 119–41).

Since the Sarit administration, socio-economic development has been employed as an instrument of national integration. It was believed that improving the socio-economic welfare of the people would reduce the level of social conflict and lessen the 'social distance' between the central government and the peripheries. In the case of the Malay-Muslims, however, this policy was complicated by religious and cultural differences: Muslims perceived the government's development efforts as an intrusion of colonial power which threatened their identity and socio-cultural values (Pitsuwan, 1982: 167–8).

Since the mid-1960s, the main thrust of the new integration strategy has been on education: government expenditures for primary education in the three southern provinces where little Thai is spoken (Thai is fluently spoken in Satun) increased at an average annual rate of 40 per cent during 1972–6 (M. Ladd Thomas, 1982: 171). Nearly every village now has a four-year primary school and village kindergartens are rapidly being established to provide a foundation in the Thai language for Muslim children. Secondary schools, which are divided into lower and higher sequences, have also received attention. Each district will soon have one secondary school of lower sequence. Vocational secondary schools have been built in each provincial capital. Teacher-training colleges as well as technical schools in the area are expanding their facilities. A university, Prince of Songkhla University, has been established with one campus in Patani Province. There has also been a steady increase in the number of adult education programmes. The General Education Centre in the provincial capital of Yala has helped to improve primary and secondary schools in the Muslim provinces; it has been experimenting with new methods of teaching the Thai language to the Muslim students and has designed teaching techniques and textbook materials, and held various conferences on ways of resolving difficulties in teaching Muslim students (M. Ladd Thomas, 1982: 171–2).

Through Thai education, the government hopes to accomplish two objectives: to achieve greater communication and understanding between Muslims and Thai government officials and to create a body of Muslims in the Thai bureaucracy. The ultimate aim is to make the Muslim population accept the authority of the Thai state. 'With the Malays themselves serving the state power structure,' noted Pitsuwan (1982: 174), 'no matter how little power they actually hold in it, the appearance of being represented and of self-government would help narrow the gap between the Thai rulers and the Malay ruled.'

The Malay traditional system of education, the *pondok* system, has gradually been transformed into a system of government-sponsored private religious schools with a secular curriculum and Thai language as the medium of instruction (for an account of the *pondok* and its transformation, see Pitsuwan, 1982: 175–204). Such transformation is regarded by the Muslims as a frontal attack on their Islamic institution, which has caused its religious quality to decline. To some Muslims, this is a serious threat to the Muslim community. While a growing number of Malay-Muslims are accepting Thai education, others feel obliged to send their children abroad for religious education.

The transformed *pondok* has not only reduced the quality of religious education; it also has a secular curriculum inferior to that of the purely secular school. Not surprisingly, many Muslim villagers who have graduated from the converted *pondok* have failed to qualify for higher education. Consequently, in 1970 the government decided to set a quota for Malay-Muslim student admissions to the nation's institutes of higher learning, without having to go through the normal entrance examinations. Through this quota programme, hundreds of young Malays have graduated from universities and other institutions and have since been recruited by the Thai bureaucracy. These Thai-educated Muslims have gradually been integrated into the Thai nation and are referred to by the Thais as 'Thai-Muslims'. They become the counterbalance to the foreign-educated Muslims who have been leading the community in other directions.

However, Thai-Buddhists and other minority groups are not happy with the special privilege accorded to the Malays. They consider that higher education, which is a 'passport' to a better life, should not favour any specific group, and view the quota programme as a 'reverse discrimination'. The strongest criticism of the programme has come from former Prime Minister Kukrit Pramoj himself. During his administration (1975–6), Kukrit proposed to abolish the whole scheme. He argued that the special privilege granted to the Muslims would cause an inferiority complex to develop among them. Their peers would regard them as being unqualified students admitted to the institutions without the rigorous entrance examinations. Some would turn out to be weak students and be forced to drop out; psychologically, they would suffer even more, and in the end would join the opposition (Suthasasna, 1976: 164). Kukrit also opposed government interference in *pondok* education. He preferred improved secular education for Malay-Muslim students. Kukrit, however, lasted only ten months as prime minister and was unable to implement his education policy; so the programme continues.

Closely related to educational measures was a decision of the central government to recruit Malay-Muslims into the bureaucracy. By doing so, the government sought to achieve the closer co-operation between Muslim villagers and government officials deemed necessary for the success of its integration efforts. However, the Muslim officials are caught between two contradictory demands: the demand of the bureaucracy for their assimilation and total loyalty to the Buddhist-oriented state, and that of their own people who expect better treatment from them. Once they change their Malay names and adjust themselves to fit Thai officialdom, they are viewed as having rejected their own culture and lose the trust of their own people (M. Ladd Thomas, 1975: 5; Suhrke, 1977: 241; Pitsuwan, 1982: 210-11). On the other hand, if Muslim officials develop close relationships with the people of their constituencies and refuse to bow to Thai bureaucratic pressure, they are labelled disloyal and their future in the bureaucracy is at stake. Thus, while increasing numbers of Malay-Muslims have been recruited into the bureaucracy, they have done very little to bridge the gap between the Muslim villagers and the state.

Along with the recruitment of Muslims into the bureaucracy, attempts have been made to promote closer relations between Muslim leadership and the central government. The hierarchical structure of the official Islamic institutions, i.e. the National and Provincial Councils for Islamic Affairs and the Council for Mosque (Royal Decrees of 1945, 1947, and 1948), reflect such attempts. The Chularajmontri or State Councillor for Islamic Affairs, who is the ex-officio head of the National Council for Islamic Affairs, is elected by a 25-member committee of the National Council and 26 Presidents of the Provincial Councils and then appointed by the King. The Chularajmontri serves as an adviser on Islamic matters to the Ministry of the Interior and the Ministry of Education as well as being a religious head of the Muslim community who issues *fatwa* to the Muslims. The position of Chularajmontri, which is considered equivalent to the Buddhist Patriarch of Thailand, is highly respected by Muslims in Bangkok and the Central Plain. In contrast, the Malay-Muslims in the south respect the Chularajmontri only in his own private capacity as a religious leader. They do not view him as a leader who has their interests to protect and promote. In fact, the Chularajmontri was refused the customary reception ceremony by the Malay-Muslim leaders on his visit to Patani during the Muslim demonstration against the government in December 1975 (To' Mina, 1980-1: 32; Pitsuwan, 1982: 212).

More significantly, the religious rulings issued by the Office of the Chularajmontri in Bangkok have little effect on the Malay-Muslims in the four provinces, who seek religious advice and *fatwa* from religious teachers and learned men in their own community. Besides, some Malay villagers feel that religious advice and *fatwa* given by the Office of the Chularajmontri might not be completely without the influence of the Thai authority. The extent of the villagers' preference for the *fatwa* of their own religious teachers is illustrated by the fact that some Malay-Muslims are always reluctant to observe *hari raya* on the day announced by the Office in Bangkok unless it conforms with the opinion of the religious teachers in their community. In 1981, for example, about half of the Malay-Muslim population in the four provinces observed their Hari Raya Aidil Adha a day earlier than that officially announced (Che Man, 1983: 89).

Several other schemes designed to secure the allegiance of the Malay villagers have been devised. At the end of 1960s, the Ministry of the Interior initiated an intensive programme of Islamic preaching called *thammacharik* (pilgrims of faith), following the Buddhist model of *thammathut* (ambassadors of faith) (Pitsuwan, 1982: 213). The *thammacharik* programme involved several members from the National Council for Islamic Affairs and religious and other leaders from Bangkok and (a few) from the Muslim provinces. Most of the *thammacharik* preachers could not speak the local Malay language, and the content of their preaching was also seen by most villagers to be very basic; the programme was ridiculed by the villagers as 'selling coconuts to the coconut farmers'. More seriously, some Muslims alleged that the *thammacharik* was intended to undermine Islam: they detected many serious errors in the Thai translation of the Quran distributed to villagers by the *thammacharik* delegation and ordered villagers to destroy them. The *thammacharik* programme was stopped, having failed to achieve its objective.

Another scheme devised by the government to improve relations with Muslim community leaders is the 'study tours' programme. Since the mid-1960s, groups of Malay-Muslim leaders from the four provinces, including religious teachers, members of the local government Islamic institutions, and other community leaders, have been selected to visit Bangkok for 'study tours' at the government's expense. During their visit, their opinions are elicited in meetings with various government leaders (Haemindra, 1977: 104). By 1985 more than 3,000 Muslim community leaders, mostly *imam* and religious teachers, had visited Bangkok under this programme (interviews with members of the Provincial Council for Islamic Affairs, Narathiwat, June 1985).

Apart from the above programmes, it has been the policy of the Thai government to emphasize the acceptance of religious plurality. The Thai monarch must be the patron of all religions (*sasanupathampok*) (*Constitution of Thailand 1978*, Article VII). Under this policy, which can be seen as a measure aimed at preventing the exploitation of religious issues by the separatist fronts, the central government has sponsored the building of mosques, especially the central mosques of the four provinces. The mosque, as discussed in Chapter 2, is the centre of all activities of the *ummah* and it stands as the most visible symbol of Islam. However, Thai government-sponsored mosques in the Muslim region are viewed by some Muslim critics as belonging to the category of Ibn Ubaiya's mosque.[1] Many Muslim villagers hesitate to pray in the central mosques built for them by the central government for the same reason. In Narathiwat, for instance, many Muslims prefer to pray in the old Al-Jamik Mosque, while in Patani they favour Chabangtiga Mosque. In Yala, the newly constructed central mosque was designed rather differently by the government architects: the main dome of the mosque is in the form of a lotus bud, a Buddhist symbol of the heart; the front stairs of the mosque resemble the structural characteristic of a Buddhist temple. Such symbols are perceived by Muslims as a sign of the government's heavy-handedness rather than its respect for Islam. As Pitsuwan (1985: 8–9) writes, 'The construction of the Yala mosque has back fired. It has not served to create an atmosphere of mutual respect and friendship. It has generated a communal tension and mutual suspicion in that province and beyond.'

Based on the same principle of religious patronage, Thai authorities exploit religious and cultural activities of the Malay-Muslims to promote closer relations between Muslim leaders and government officials. For example, the Maulud (Prophet Muhammad's birthday celebration), which was normally celebrated by Muslims at the family and community level, has been upgraded by the Thai provincial authorities to an annual festival of the provinces. The Governors and the District Officers usually become hosts of the Maulud ceremonies. Some Muslim leaders, however, interpret the government's involvement as an attempt to manipulate their religious activities for political purposes and believe that the Maulud celebration is being spoiled and belittled by the participation of non-Muslims. Indeed, Islam does not approve the involvement of non-believers in the performance of Islamic pious duties.

With respect to economic measures, there are nation-wide economic development programmes directed towards improving economic conditions throughout Thailand. There are no programmes exclusive to the Muslim provinces, and economic programmes in the four provinces have often received less attention and funding from Bangkok than those of other regions. However, the construction of roads and replanting of rubber seem to be greater in the Muslim areas. Roads in the Muslim provinces had long been either non-existent or very poor compared to the network of roads across the border in Malaysia. Malay-Muslims were prone to view this as the most concrete evidence of Bangkok's neglect of the area. In the mid-1960s, the government began to allocate funds for road construction in the border provinces. With financial and technical assistance from abroad, the government has been able to build and repair roads throughout the region (M. Ladd Thomas, 1982: 166). National and provincial highways are now paved and many other main roads have been either paved or given a laterite surface. One can now drive, with only a few exceptions, between district centres throughout the provinces whatever the season.

Most of Thailand's rubber plantations are in the Muslim provinces. The rubber replanting scheme, which was also initiated in the mid-1960s, aims at the eventual replacement of most rubber trees with more productive strains to increase the earnings of the villagers. The government subsidizes plantation owners who replace old rubber trees with improved strains. The subsidy covers the major costs of clearing out old trees and replanting; it does not compensate for loss of earnings while the new trees reach their production stage. Funds for this purpose come from a special tax on rubber exports and a loan of about US$50 million from the World Bank (M. Ladd Thomas, 1982: 169). Plantation owners who want to participate in the programme apply to the Rubber Replantation Board in the provinces. They must indicate the area of land they wish to cultivate. The Rubber Replantation Board at the provincial level administers the payment of subsidies, but each application must be approved by the Board in Bangkok. Experts at the Rubber Research Centre in Had Yai estimated that a plantation owner who replanted with the best strains and followed the most modern methods of processing rubber would be able to earn fourteen times as much per tree. Yet, many Muslim rubber smallholders hesitate to seize the opportunity. One reason is that for the smallholders, who are mostly Malays, it is financially not feasible to cut down old rubber trees and lose their only source of income during the replanting period. Secondly, bureaucratic red tape in processing applications makes Malay Muslims who cannot speak Thai reluctant to participate. Thirdly, many Malay-Muslims do not have clear title to their land. They fear that the government might raise the issue of legal ownership of their land, or that they might have to pay an expensive fee for land survey if they apply for a subsidy (M. Ladd Thomas, 1982: 170). Finally, the usual mistrust between the Malay villagers and Thai officials causes this very ambitious scheme to fall short of its objective.

Efforts have also been made to correct the inadequacies of administration in the Muslim provinces. The central government has taken greater care in choosing Thai officials for the region. Some Malay-language training and limited orientation in Malay customs and religion are given to Thai officials. Furthermore, in the late 1970s the government authorized General Harn Leenanon, then Commander of the Fourth Army Region, to remove corrupt senior officials, known as 'dark influence', in the south (Leenanon, 1984: 25–39; Snitwongse, 1985: 265). Several regional centres were created in Yala Province to improve administrative efficiency, such as the Centre for the Administration of the Southern Border Provinces, the Community Development Centre, and the General Educational Centre. Nevertheless, corruption and other administrative problems remain unsolved. There is a feeling that the government needs to involve more local people in helping to deal with the problem.

Finally, the Thai government has responded to the Muslim separatist problem by suppression. As discussed in Chapter 3, large-scale military and police operations were carried out against the various fronts as they increased their guerrilla activities at the end of the 1960s. This heavy-handed military response, which lasted seven years from 1968 to 1975 (Megarat, 1977: 21–2), created more grievances among the Malay-Muslim population. The largest demonstration in Patani in 1975 caused the Thai government to re-evaluate its policy (Snitwongse, 1985: 266) and in 1980 it initiated its first long-term 'policy of attraction'. The result was that some 450 separatists from different fronts surrendered during the period September 1982 to October 1983 (*The Star*, 7 October and 3 November 1983). Along with the policy of attraction, diplomatic representations have been made to OIC member states to discourage certain countries from providing assistance as well as to try to keep the movements from having representation in the international Islamic forums. The various forms of co-operation with the government of Malaysia, particularly in suppressing the CPM (see *Bangkok Post*, 18 March and 1 May 1987), can be seen partly as an attempt to gain a quid pro quo in its refraining from assisting the Muslim separatist fronts.

Concluding Remarks

The involvement of external parties in the Malay-Muslims' struggle against Thai domination since the beginning of the twentieth century has helped to raise their consciousness and produced demands for autonomy and independence. This stimulus, consisting of expressions of sympathy and material assistance, has come principally from neighbouring Malaysia. The concern of the Malaysian government for the welfare of the Malay-Muslims in Patani helped to restrain the Thai government from employing a rigid assimilationist policy or sharpening its military suppression of Muslims. But external support from Malaysia as well as from other Muslim countries has been insufficient to achieve the major objective of the separatists. The Malaysian government could not commit itself to total support of the struggle without jeopardizing its national interests. The purpose of Malaysian involvement, then, has been to prevent the Thai government from taking action that might yield results undesirable to Malaysia and to its ethnic brethren across the border.

Along with military operations, the Thai government has since the mid-1960s initiated socio-economic and political development programmes aimed at preventing the separatist movement from gaining wider support within the Muslim community. Among these programmes, two areas have been emphasized: economic development and Thai education. The assumption is that the former is necessary to stop the economic deterioration of the Muslim region which is susceptible to separatist arguments. The latter is crucial for political integration of the Malay-Muslim community. However, few of the economic programmes have been carried out with a commitment consistent with Bangkok's expressions of concern. The economic problems of the rural Muslims remain unsolved. On the other hand, the education programmes have proved to be successful. Most of the new-generation Malay children and youths attend schools that offer a Thai secular curriculum. Their numbers in Thai secondary and tertiary institutions have increased markedly. And the traditional *pondok* system of education has been almost completely destroyed.

It remains to be seen, however, whether the government's successful penetration of education can be translated into greater loyalty to and identification with the Thai nation. If the economy and bureaucracy cannot absorb these Thai-educated Malay-Muslims, they might well be recruited by the separatist leadership.

External Influences on and Government Responses to the Moro and the Malay Separatist Conflicts Compared

The external factors that affect the Moro and the Malay struggles are in some aspects difficult to distinguish. The resurgence of Islam, for instance, helps to raise the consciousness of both groups as distinct communities. The involvement of certain Muslim states and organizations serves to strengthen the demands for autonomy and independence and to help curb the actions of the Philippine and Thai governments against the minority populations. In the case of Thailand, the expectation of a strong reaction from Malaysian and other Muslim governments has restrained Bangkok from pursuing a policy of suppression; in the Philippines, the intervention of the OIC has restrained Manila in its conduct of the war against the MNLF (Suhrke and Noble, 1977: 209).

External support for both movements is based mainly on ethnic affinity or Islamic brotherhood. The Moros in Mindanao and the Malays in Patani are parts of the 200 million people of the Malay race in South-East Asia, and the Islamic bond links them with the wider world of Islam. The ultimate aim of the external supporters of the two movements (apart from Malaysian assistance to the Moros in direct reaction to the Philippine claim to Sabah) is essentially to prevent actions that might result in the destruction of these Muslim communities, as occurred historically to the Muslims in the Iberian Peninsula (711–1614) and in Sicily (827–1246) (see Kettani, 1986). Nevertheless, they do not support the separatist objective of the movements which undermines the sovereignty of the nation-state.

The major difference between the Moro and the Malay struggles lies in the extent of the conflict, which has been more vicious and intense in the Philippines than in Thailand. This is in part the reason why Muslim states and organizations have been more concerned with and involved in the Moro conflict. Indonesia, the most populous Malay state in the region, for example, defends its non-involvement in the Patani Malay separatist struggle by denying the validity of ethnic self-determination as a general principle; but, in a limited capacity, it has been involved in the Moro struggle.

The Philippine and Thai governments respond to their respective separatist threats by employing socio-economic measures along with military suppressive operations. Both Manila and Bangkok have accelerated programmes of socio-economic and educational upliftment in the Muslim regions. Special provisions for the partial application of Islamic law for Muslims are provided in both countries. Attempts have been made by both governments to deal with shortcomings in the local administration by orientation-training for non-Muslim government officials assigned for duty in Muslim areas.

However, the governments' measures in both cases appear to have come partly in response to external pressures and partly from a desperate desire to restore order in deteriorating situations rather than from a genuine respect for and understanding of the Muslim *ummah*. In other words, the negative attitudes of the Philippine and Thai governments have prevented them from responding adequately to the needs and aspirations of the Muslim minorities, in turn limiting their success at penetrating the Muslim communities.

In the final analysis, the persistence of the separatist movements in the southern regions of the Philippines and Thailand will depend largely on the involvement of Muslim states and on the manner in which Manila and Bangkok respond to the separatists. While the governments regard cultural autonomy as a threat to national integrity, Muslim identity has evolved to a point where rectification of socio-economic ills is no longer an adequate method of accommodation. Unless some sort of meaningful 'cultural autonomy' is granted to the Muslims of Mindanao and Patani, the separatist struggles in the two countries will persist.

1. In the days of the Prophet Muhammad (S.A.W.), Abdullah Ibn Ubaiya built a mosque just to show off; the Prophet ordered it to be destroyed because it was a product of hypocrisy rather than of real faith.

6
Conclusion

IN the introductory chapter, the persistence of ethnicity was noted. This has been viewed not only in the context of a broader ethnic revival, which, under the banner of national self-determination, has been developing since the late eigthteenth century, but also as the recent phase of a long historical cycle of ethnic resurgence and decline which has been occurring since the dawn of recorded history (Smith, 1982: 86; cf. Calvert, 1970). Various explanations have been suggested for the persistence of the ethnic community. The central argument, however, has been that the ethnic community has long been a socio-cultural 'pattern' for human organization and communication. It holds in common a set of traditions not shared by others. Such traditions include common myths of descent or place of origin, distinct cultural practices, and a sense of historical continuity. These 'primordial' ties produce a strong sense of ethnic identity and solidarity, which in time of threat or outside pressure can override class, faction, and other divisions within the community. At times ethnic cohesion may be overlaid by other types of allegiance, but from time to time, particularly in the face of external encroachments, it will re-emerge in sufficient force to override other kinds of loyalties.

Three types of ethnic minority community are distinguishable: minority immigrant populations, normally a by-product of labour-recruiting policies during a colonial era; indigenous peoples who have become a minority as an outcome of a colonial settlement; and minority groups which result from incorporating hitherto autonomous peoples or tribes under an alien authority. This study has examined separatist movements of the Moros and the Malays, the third category of minority communities, who regard separatism as a political possibility because of a socio-historical logic, the coincidence of geography and cultural plurality, distance from the centre of authority, and the support of an external community.

This study has traced the origins and evolution of Muslim sultanates in Mindanao and Patani, as a root cause of centuries of resistance struggles against foreign domination. The sultanates were also the foundation of social and political structures of contemporary Moro and Malay societies, where Islamic principles and pre-Islamic Malay traditions and customs remain strong. Both communities perceived their

CONCLUSION

resistance efforts as both political and religious, and political and religious leaders were often involved. More importantly, both leadership groups saw their involvement as a religious obligation, to comply with the teachings of the Quran and Hadith; and thus Islam became the primary rationale for their struggles. But the ability of the two Muslim communities to resist Philippine and Thai domination was affected substantially by external intervention. Failure to prevent incorporation of their communities into the larger Philippine and Thai political systems was influenced, to a great extent, by the intervention of the Western colonial powers: without the interference of the United States in Mindanao and of Great Britain in Patani, the Moros and the Malays might not have been set upon their present course of history.

The contemporary resistance struggles began when the Moro and the Malay communities were subjected to American and Thai rule at the beginning of the twentieth century. In Mindanao, armed resistance movements, ranging from full-scale battles to minor incidents, occurred during the American and Commonwealth regimes (1899–1946), motivated by the desire to preserve the autonomy of the Muslim community. The Moro movement persisted after Moroland was structurally integrated into the Republic of the Philippines. The Moros, who constituted a nationality distinct from, and older than, that of Christian Filipinos, possessed a strong sense of group consciousness and continued to assert their identity as Muslims. As Manila pressed for national integration, Moro expectations of the benefits of participating in Philippine society were shattered by socio-economic deprivation, increasing political disadvantage, competition with Christian migrants for land, and government interference in the local affairs of Muslims. Their sense of deprivation and conflicting loyalties was further compounded by the resurgence of Islam. From the late 1960s, the Moro resistance movement began to transform from intermittent rebellions to organized liberation fronts demanding independence.

In Patani, a series of resistance movements and uprisings broke out after the region was incorporated into the Thai Kingdom at the beginning of the century. The leadership of these early rebellions comprised mainly the former royal families and religious leaders who were motivated primarily by the desire to recapture the authority of the deposed Malay sultans. From 1922 to 1947, the resistance struggle of the Malay-Muslims was largely a response to events connected with the Thai assimilationist policy. Bangkok was accused of 'Siamifying' the Malay community and of stamping out their religion and customs. Resistance and resentment against Thai efforts of assimilation were further aggravated by nationalistic aspirations among the Muslims which were stimulated by the success of independence movements elsewhere in South-East Asia. As a result, the period between 1947 and 1959 saw Patani struggle for autonomy or irredentism with Malaya. GAMPAR and PPM were organized to carry out such struggle. This period also witnessed the transfer of leadership from former aristocrats to religious leaders. From the late 1950s, the Thai government moved aggressively to consolidate its control over the socio-economic and religious affairs of the Malay-Muslims. Especially sensitive was government interference in religious institutions such as the *pondok*, at a time when Islamic awareness among the Malay-Muslims was increasing with the general resurgence of Islam. In the 1970s, the three major liberation fronts flourished.

The outbreak of separatist conflict in Mindanao and Patani in the 1970s was but the continuation of the centuries-old history of Muslim struggles against foreign rule. The fact that the Moros and the Malays are conscious of their historically and culturally coherent communities as sovereign states has provided them with the legitimate principle and inspiration to carry on their separatist struggles. Moreover, the movements have enjoyed persistent local support. In the process of nation-building, certain minority groups become anomalies, without the same trust or benefits enjoyed by members of the national majority. As McVey (1984: 12) has explained, 'Nationality is conceived in the modern nation-state as a kind of super-ethnicity ... around which a national cultural boundary is created, with minority groups tending to fall outside.' Filipino nationalism, for example, regards Christianity as an essential ingredient; Muslims, therefore, are an anomaly.

As mentioned above, more particular factors triggered the separatist conflicts. First, the government's acceleration of integration attempts and the specific cases of maltreatment that characterized the 'internal colonialism' of the socio-economic and political situation of the Muslim communities deepened the sense of deprivation and broadened ethnic solidarity. Hannan (1979) has observed that the penetration of the centre, through economic and political developments, creates the conditions under which ethnicity is considered as the most viable and effective political instrument for the periphery to oppose the centre. And Hechter (1975) has noted that the extent of overlap between minority and low-reward occupations produces ethnic solidarity. In other words, modernization, in the form of increases in social mobilization, intensifies ethnic tension and fosters ethnic competition which can be conducive to separatist demands.

Secondly, the general resurgence of Islam since the end of the Second World War has raised Islamic consciousness among the Moros and the Malays. It has widened the gap between Moros and Filipino-Christians in the Philippines and between Malays and Thai-Buddhists in Thailand. At the same time, it has bridged the relations of Moros and Malays with the outside Muslim world and thus strengthened their identity as Muslims. More importantly, Islam, as a political ideology, has been used to provide legitimacy and inspiration and for mass mobilization. Both separatist movements are, therefore, legitimated and inspired in religious terms; all the existing liberation fronts have employed the same motto—*Allahu Akbar!* (God is most great!) Through its institutions and rituals such as mosque and Friday prayers, Islam mobilizes people for collective action. Throughout the history of Islam, as recently in the Iranian Revolution, the mosque has provided a place for people to assemble and to discuss grievances and revolutionary ideas. The *imam* and other mosque functionaries often become a source of active opposition to secular rules (Siddiqui, 1982: 142).

Together with the general resurgence of Islam, the increase in oil prices in the 1970s brought a growth in the economic power of the Arab world. As a result, Islamic institutions such as the Organization of the Islamic Conference, the Islamic Development Bank, the Islamic Solidarity Fund, the Muslim World League, the Muslim World Congress, and the World Islamic Call Society of Libya, increased their activities throughout the Muslim world in an attempt to preserve and strengthen the *ummah* (Kettani, 1986). These institutions have assisted or influenced in varying degrees the struggles of Muslims in Mindanao and in Patani.

A question remains as to whether the emergence and expansion of the Moro and the Malay separatist movements in the 1970s were primarily attributable to economic exploitation, as suggested by some writers, or whether they were mainly a response to cultural degradation. Relative economic deprivation has been an element of the situation of the Muslim communities in the southern Philippines and southern Thailand, but what seems to be at issue primarily is a conflict of cultures which is seen as the continuation of centuries of confrontation between Muslims and foreign intrusion. Moreover, since there is no clear distinction between economics, politics, and culture in the eyes of Muslims, the two separatist conflicts are sustained by the belief that the continued efforts of the centres to consolidate their control over the socio-economic and cultural affairs of the Muslim communities, will 'eventually lead to the erosion of their cultural and religious life and the disappearance of their communities' (Abubakar, 1984: 78; cf. Kettani, 1986).

Certain notable features of the Moro and the Malay separatist movements can be observed. In the case of a conventional revolutionary front, such as the NLF, the effectiveness of the organization is measured in part by the degree of institutionalization of its organizational machinery, including a clearly established chain of command. The Muslim movements, however, consist of loosely organized liberation fronts; their strength depends less on conventional principles of effective organization than on the fact that they are ethnically based, religiously motivated, and led by élite groups that dominate their respective communities. The MNLF and the BNPP, for instance, are loosely knit fronts, unable to control the behaviour of the various factions or to establish a clear chain of command. The fronts' policy-making bodies, the central committees, satisfy themselves with furnishing broad policy outlines, giving the local leaders the power to make their own decisions. There are no specific criteria of membership; admissions are accompanied by no official formalities. Only the adherent's activity within the fronts can determine the degree of participation. One cannot, therefore, measure the strength of the movements by simply counting the number of existing armed guerrilla units or present clashes with government forces. The strength and activities of this kind of underground movements are difficult to predict; they depend on various internal and external situations and stimuli that can either trigger or retard their development. The Jabidah Massacre that led to the emergence of the MNLF in the Philippines and the Patani Massacre that triggered the largest and longest Muslim demonstration against the Thai government are cases in point.

Also, the movements are characterized by broadly defined aims. Depending on internal and external circumstances, the objectives of the fronts change from independence to autonomy or irredentism or qualified participation. This suggests that the movements do not actually seek national sovereignty, but rather respond to integration attempts by the centre. The extent of their response often corresponds to the degree of repressiveness of government policies. In a sense, the movements essentially represent struggles for survival, for the preservation of Muslim identity and way of life.

In addition to its role as a mobilizing force and in legitimizing revolutionary action, Islam is also regarded as the ideological foundation of the movements. But although all major fronts are founded on Islamic ideology, each has its own coloration. Despite the bond of Islam, the movements have become factionalized. In the case of the Moros, ethno-cultural divisions, differences in ideological inclination, and family backgrounds have been the primary cause of factional cleavages. Even though the Malays have no ethno-cultural splits, their factional battles derive mainly from conflicting ideological preferences and educational and family backgrounds. Disunity among the Muslim states further reinforces factional conflicts in both movements. Factionalism has, indeed, become one of the dominant features of ethnic-minority separatist struggles. Apart from its negative side, factionalism, however, seems to have helped to keep the conflicts alive; for instance, the refusal of the MILF to give up its armed resistance against the Aquino government, while the MNLF has agreed to negotiate for autonomy, has kept the movement in Mindanao going.

Another notable feature of the Moro and the Malay movements is that they are organized and led by dominant leadership groups in their respective communities. They are, therefore, not insignificant and isolated affairs of the communities. The traditional aristocratic élite, which dominates the Moro community, has played a major leadership role. Of the four separatist groups, only the Nur Misuari Faction of the MNLF has been led by the secular élite group; the Salamat Hashim and Yusoph Lucman groups are dominated by the traditional and religious élites, whereas the Dimas Pundato Faction of the MNLF is largely controlled by traditional and secular leaders. In the case of the Malays, the four existing fronts have been led primarily by religious leaders who are the influential élite group in Patani society. Other Muslim leadership groups in Mindanao and Patani have also been involved in various capacities in the separatist struggles to protect their identity as the *ummah*.

The fact that the Moros and the Malays are a part of the Muslim *ummah*, and of the Malay world which constitutes the largest group of Muslim population of a single race, renders the movements even more significant. The involvement of certain Islamic institutions and Muslim states in these struggles was based precisely on the *ummah* identification and ethnic affiliation with a view to safeguarding dignity and national rights and to relieve the persecution of the Muslims. However, external participation was also motivated by the national interest of the individual state. Malaysia's involvement in the Moro struggle, for instance, was a direct reaction to the Philippines' claim to Sabah and a consequence of the Islamic fervour of Tun Datu Mustapha. External assistance helped to strengthen the military and political capabilities of the Moro movement to the extent that it was able to put pressure on Manila to make some concessions and restrain its more repressive policies. External support from Malaysia and other Muslim communities to the Malays in Patani was not sufficient to achieve the main objectives of the separatists, but it restrained the Thai government from intensifying its military suppression and assimilationist policy.

Despite these similarities, the Moro and Malay movements differ in a number of respects. Differences in leadership and ethno-cultural composition have been discussed earlier, but the fundamental difference is that the Moro movement was able to stage a more serious challenge to Manila than the Patani movement was able to pose to Bangkok. There are several reasons for this. While repression was pursued aggressively by the Philippines during the Marcos era, the Thai assimilation policies toward the Malays in Patani during the same period were restrained by anticipated strong reaction from the neighbouring Malaysian government. More specifically, a greater resistance by the Moros was mainly due to two factors: first, Christian migration, resulting in shifts in economic and political power from Muslims to Christian predominance; and secondly, military involvement in violence resulting from the first factor. Such repressive policies of the Philippines not only led to equally strong response from the Moro people but also attracted involvement and sympathy from the overseas Muslim community. This external factor helped to intensify the Moro conflict to the extent that a protracted war forced the Marcos government to sign the Tripoli Agreement. Secondly, the Moro struggle, which was led by different categories of Muslim leadership—traditional, secular, and religious—benefited from having strong charismatic leaders who were able to develop effective international contacts. In contrast, the Malay movement, which was dominated primarily by the religious élite, appeared to be less willing or less able to expand its external contacts and lacked charismatic leaders capable of providing a serious threat to the Thai government.

As a consequence, the two movements obtained different degrees of success. The Moros succeeded in getting some concessions—whether real or nominal—from Manila. These included the establishment of Autonomous Regions IX and XII, the formulation of a number of development programmes in Mindanao and the recognition of some customary aspects of Moro society. The constitution of 1987, moreover, makes provision for the creation of a Muslim autonomous region. These concessions suggest that the Philippine government has finally recognized the Moro people as a historically and culturally distinct community. Nevertheless, the basic goal of the struggle, to gain meaningful autonomy, remains unrealized. The Malay movement in Patani, on the other hand, has received neither direct concessions nor recognition from Bangkok, though its separatist activities probably helped to restrain the Thai government from pursuing coercive policies of assimilation.

Even a superficial examination of separatist movements elsewhere in the world suggests a number of common features between the Moro and Patani movements and other ethnic separatist movements: in their persistence; their tendency to respond to perceived threats from centralizing states or dominant ethnic groups; the looseness of their organizational structures and ideologies, and even their apparent tendency to factionalize, while at the same time commanding the often unarticulated loyalty and potential political and military threat of ethnic mobilization. What perhaps distinguishes these Muslim separatist movements from the rest is that they have integrated Islamic concepts and symbols into a national dogma, and that Islam and nationalism reinforce each other in their struggles against foreign rulers, while at the same time linking them with the wider Islamic *ummah*.

No multi-ethnic state has proven immune to the surge of ethnicity. Authoritarian, democratic, federative, and unitary have all been affected. Nor does the proliferation of international organizations and alliances decrease the significance of ethnic nationalism. On the contrary, the post-war international economic and political organizations, which emphasize membership, negotiation, and power only among nation-states, appear to have encouraged minority communities to think in nationalistic terms (Connor, 1967, 1969, 1972; Birch, 1978). More importantly, no government of a multi-ethnic state has found the solution to the problem posed by the demands for modernization on the one hand and the tendencies of growing ethnic nationalism on the other. Determined to prevent secession and to achieve national integration, many governments tend to resist separatist movements with coercive methods, while simultaneously promoting assimilation. Such policies have proved remarkably unsuccessful.

Thus, while some Moro and Malay élites are gradually absorbed into the Philippine and Thai systems through the process of national integration and development, policies which seek to redress the separatist problem through socio-economic measures designed to lift the living standards of the ethnic minorities fail to recognize that the ethnic protagonists perceive their conflict not in socio-economic terms but as ethnic, religious, and nationalist.

Muslim armed separatist movements in the Philippines and Thailand are likely to persist in the future. As historically autonomous and distinct peoples, the resistance of the Moros and the Malay-Muslims will be directed at government policies of assimilation and bureaucratic exploitation or maltreatment. The persistence of the movements will reflect their will to survive, and their struggles will likely be characterized by periodic resurgence and recedence depending on the internal and external factors which trigger or retard them. There is no reason to expect their secular decline. While Manila and Bangkok consider cultural autonomy for the Muslim communities to be a threat to national territorial integrity, the Moros and the Malays regard the concept of 'national self-determination' as a fundamental right of every people, believing, with Woodrow Wilson (quoted in Connor, 1967: 31), that 'every people has a right to choose the sovereignty under which they shall live'.

Glossary

adat	customary laws, customary practices
agama	religion, socio-religious community
akal	rationality
alim	learned man, religious learned man
anting-anting	charms
balaisa	prayer-house
bangsa	nation, identification with ancestors
baniaga	slave
barakat	charismatic grace
barangay	community, boat
bendahara	prime minister
bendahari	treasurer
berkat	blessed
bilal	prayer caller
bintala	district supervisor of priests
boriween chet huamuang	Area of the Seven Provinces
Bunga Mas	Flowers of Gold
cedula	head tax
chaomuang	head of state, local ruler
Chularajmontri	State Councillor for Islamic Affairs
conquistadores	conquerors
cota	fort
daerah	territorial district
dakwah	call of faith
dar al-Islam	territory of Islam
datu	traditional leader, man of rank
deen wa dawla	religion and socio-political order
du-ah	Muslim preacher, Muslim missionary
dumatu	lesser noble
fatwa	religious ruling
guru	teacher, religious teacher
Hadith	Prophetic traditions
haj	pilgrimage to Mecca
hajji	Muslim who has fulfilled pilgrimage duty
hamba	slave
hari raya	Muslim festive day

ibadat	pious duties
imam	prayer leader
jihad	exertion, holy war
kafir	infidel, non-believer
kafir al-harb	non-believer of the first category
kamnan	commune headman
kampilan	sword
kampung	village
kenduri	communal feast
khaek	alien, visitor
khaek chet huamuang	Seven Malay Provinces
khaluang boriween	Area Commissioner
khatib	sermon's conveyor, preacher
khru	schoolteacher
khutba	sermon
kuffar	infidels
laksamana	admiral
laksmanna	*panglima*'s messenger
Lupong Tagapagpaganap Ng Pook	Executive Council
madaris	religious schools
madrasa	religious school
mahatthai	Ministry of the Interior
Majlis Agama Islam	Provincial Council for Islamic Affairs
Majlis Bechara	Assembly of Discussion
Majlis Ugama Islam	Islamic Religious Council
maratabat	honour
masjid	mosque
Maulud	occasion of commemorating Prophet Muhammad's birthday
monthon	circle
muazzin	prayer caller
mudarris	teacher
mufti	chief judge
mujahideen	participants in the holy war; Muslim fighters
nafsu	irrationality
nakib	officer in charge of military matters
nakuda	leader of trading expedition, successful trader
negeri	state
nikhom sangton-eng	Self-help Colony
orang alim	learned man, religious learned man
orang berhutang	debt-bondsman
orang beriman	pious man
orang kaya	rich man, successful trader
orang sakai	aborigines
panglima	sultan's personal representative, district chief

parkasa	*panglima*'s aide-de-camp
patthanakorn	community development officer
pendita	learned man, religious learned man
perahu kolek	small fishing boat
perlembagaan	constitution
phuyaiban	village headman
pondok	Islamic religious school, traditional Islamic school
prenda	mortgage
pusaka	inherited property, inheritance
qadi	Muslim judge
Quran	Holy Book of Muslims
ra'ayat, rakyat	subjects
raja	ruler
rasa	race
riba	interest
rongrian aekachon sornsasana Islam	private Islamic school
rongrian sonsasana Islam	Islamic religious school
Rumah Bechara	House of Discussion
sabilillah	struggle in the name of Allah
sakop	follower
salat	prayer
Sangguniang Pampook	Legislative Assembly
sasanupathampok	patron of all religions
semangat	zeal
shahbandar	harbour master
sharia	Islamic law
shura	Islamic-oriented meeting
sultan	traditional leader, ruler
tanah Melayu	lands of the Malays
tarsila	genealogy, traditional literature
tau way bangsa	commoners
temenggong	minister of war
Thai ratthaniyom	Thai Custom Decree
thammacharik	pilgrims of faith
thammathut	ambassador of faith
thesaphiban	administrative centralization system
timuway	headman
tok guru	traditional religious teacher
to'na	master
ulama	learned men, religious learned men
ulipun	debt-bondsman
ummah	Muslim community
ustaz	non-traditional religious teacher
zakat	alms

Appendices

APPENDIX 1

The Tripoli Agreement

IN THE NAME OF GOD, THE OMNIPOTENT, THE MERCIFUL.

AGREEMENT BETWEEN THE GOVERNMENT OF THE REPUBLIC OF THE PHILIPPINES AND THE MORO NATIONAL LIBERATION FRONT WITH THE PARTICIPATION OF THE QUADRIPARTITE MINISTERIAL COMMISSION MEMBERS OF THE ISLAMIC CONFERENCE AND THE SECRETARY GENERAL OF THE ORGANIZATION OF THE ISLAMIC CONFERENCE.

In accordance with the Regulation No. 4 Para. 5 adopted by the Council of Ministers of the Islamic Conference in its Fourth Session held in Benghazi, Libyan Arab Republic during the month of Safar 1393 H. corresponding to March 1973, calling for the formation of Quadripartite Ministerial Commission representing the Libyan Arab Republic, the Kingdom of Saudi Arabia, the Republic of Senegal and the Republic of Somalia, to enter into discussions with the Government of the Republic of the Philippines concerning the situation of the Muslims in the South of the Philippines.

And in accordance with the Resolution No. (18) adopted by the Islamic Conference held in Kuala Lumpur, Malaysia in Jumada Alakhir 1393 H. corresponding to June 1974 A.D. which recommends the searching for a just and peaceful political solution to the problem of the Muslims in the South of the Philippines through negotiations.

And in accordance with the Resolution No. 12/7/S adopted by the Islamic Conference held in Istanbul in Jumada El-Ula 1396 H. corresponding to May 1976 A.D. empowering the Quadripartite Ministerial Commission and the Secretary General of the Islamic Conference to take the necessary steps for the resumption of negotiations.

And following the task undertaken by the Quadripartite Ministerial Commission and the Secretary General of the Islamic Conference and the discussions held with H. E. President Marcos, President of the Republic of the Philippines.

And in realization of the contents of Para. 6 of the joint Communiqué issued in Tripoli on the 25th Zulqeda 1396 H. corresponding to 17th November 1976 A.D. following the official visit paid by the delegation of the Government of the Philippines headed by the First Lady of the Philippines Mrs. Imelda Romualdez Marcos to the Libyan Arab Republic and which calls for the resumption of negotiations between the two parties concerned in Tripoli on the 15th of December 1976 A.D.

Negotiations were held in the City of Tripoli during the period between 24th Zulhija 1396 H. to 2nd Moharram 1397 H. corresponding to the period from 15th to 23rd December 1976 A.D. at the Ministry of Foreign Affairs presided over by Dr. Ali Abdussalam Treki, Minister of State for Foreign Affairs of the Libyan Arab Republic, and comprising the Delegations of:

1. Government of the Republic of the Philippines, led by Honorable Carmelo Z. Barbero, Undersecretary of National Defense for Civilian Relations.

2. Moro National Liberation Front, led by Mr. Nur Misuari, Chief of the Front.

And with the participation of the representatives of the Quadripartite Ministerial Commission:

The Libyan Arab Republic represented by Dr. Ali Abdussalam Treki, Minister of State for Foreign Affairs.

The Kingdom of Saudi Arabia—H.E. Salah Abdulla El-Fadl, Ambassador of the Kingdom of Saudi Arabia, Libyan Arab Republic.

The Republic of Senegal—Mr. Abubakar Othman Si, Representative of the Republic of Senegal and Charge d'Affairs of Senegal in Cairo.

Democratic Republic of Somalia—H.E. Bazi Mohamed Sufi, Ambassador of the Democratic Republic of Somalia, Libyan Arab Republic.

With the aid of H.E. Dr. Ahmad Karim Gai, Secretary General of the Organization of Islamic Conference and a delegation from the Secretariat General of the Conference composed of Mr. Qasim Zuheri, Assistant Secretary General, and Mr. Aref Ben Musa, Director of Political Department.

During these negotiations which were marked by a spirit of conciliation and understanding, it has been agreed on the following:

First: The establishment of Autonomy in the Southern Philippines within the realm of the sovereignty and territorial integrity of the Republic of the Philippines.

Second: The areas of the autonomy for the Muslims in the Southern Philippines shall comprise the following:

1 Basilan
2 Sulu
3 Tawi-Tawi
4 Zamboanga del Sur
5 Zamboanga del Norte
6 North Cotabato
7 Maguindanao
8 Sultan Kudarat
9 Lanao del Norte
10 Lanao del Sur
11 Davao del Sur
12 South Cotabato
13 Palawan
14 All the cities and villages situated in the above mentioned areas.

Third:

1. Foreign policy shall be of the competence of the Central Government of the Philippines.

2. The National Defense Affairs shall be the concern of the Central Authority provided that the arrangements for the joining of the forces of the Moro National Liberation Front with the Philippine Armed Forces be discussed later.

3. In the areas of the autonomy, the Muslims shall have the right to set up

their own Courts which implement the Islamic Shari'a laws. The Muslims shall be represented in all Courts including the Supreme Court. The representation of the Muslims in the Supreme Court shall be upon the recommendation from [sic] the authorities of the Autonomy and the Supreme Court. Decrees will be issued by the President of the Republic for their appointment taking into consideration all necessary qualifications of the candidates.

4. Authorities of the autonomy in the South of the Philippines shall have the right to set up schools, colleges and universities, provided that matters pertaining to the relationship between these educational and scientific organs and the general education system in the state shall be subject of discussion later on.

5. The Muslims shall have their own administrative system in compliance with the objectives of the autonomy and its institutions. The relationship between this administrative system and the Central administrative system to be discussed later.

6. The authorities of the autonomy in the South of the Philippines shall have their own economic and financial system. The relationship between this system and the Central economic and financial system of the State shall be discussed later.

7. The authorities of the autonomy in the South of the Philippines shall enjoy the right of representation and participation in the Central Government and in all other organs of the State. The number of representatives and ways of participation shall be fixed later.

8. Special Regional Security Forces are to be set up in the area of the autonomy for the Muslims in the South of the Philippines. The relationship between these forces and the central security forces shall be fixed later.

9. A Legislative Assembly and an Executive Council shall be formed in the areas of the autonomy for the Muslims. The setting up of the Legislative Assembly shall be constituted through a direct election, and the formation of the Executive Council shall take place through appointments by the Legislative Assembly. A decree for their formation shall be enacted by the President of the Republic respectively. The number of members of each assembly shall be determined later on.

10. Mines and mineral resources fall within the competence of the Central Government, and a reasonable percentage deriving from the revenues of the mines and minerals [shall] be fixed for the benefit of the areas of the autonomy.

11. A mixed Committee shall be composed of representatives of the Central Government of the Republic of the Philippines and the representatives of the Moro National Liberation Front. The mixed Committee shall meet in Tripoli during the period from the 5th of February to a date not later than the 3rd of March 1977. The task of the said Committee shall be charged to study in detail the points left for discussion in order to reach a solution thereof in conformity with the provisions of this agreement.

12. Ceasefire shall be declared immediately after the signature of this agreement, provided that its coming into effect should not exceed the 20th of January 1977. A Joint Committee shall be composed of the two parties with the help of the Organization of the Islamic Conference represented by the Quadripartite Ministerial Commission to supervise the implementation of the ceasefire.

The said Joint Committee shall also be charged with supervising the following:

a. A complete amnesty in the areas of the autonomy and the renunciation of all legal claims and codes resulting from events which took place in the South of the Philippines.

b. The release of all political prisoners who had relations with the events in the South of the Philippines.

c. The return of all refugees who have abandoned their areas in the South of the Philippines.

d. To guarantee the freedom of movements and meetings.

13. A joint meeting [shall] be held in Jeddah during the first week of the month of March 1977 to initial what has been concluded by the Committee referred to in Para. 11.

14. The final agreement concerning the setting up of the autonomy referred to in the first and second paragraphs shall be signed in the City of Manila, Republic of the Philippines, between the Government of the Philippines and Moro National Liberation Front, and the Islamic Conference represented by the Quadripartite Ministerial Commission and the Secretary General of the Organization of the Islamic Conference.

15. Immediately after the signature of the Agreement in Manila, a Provisional Government shall be established in the areas of the autonomy to be appointed by the President of the Philippines; and be charged with the task of preparing for the elections of the Legislative Assembly in the territories of the Autonomy; and administer the areas in accordance with the provisions of this agreement until a Government is formed by the elected Legislative Assembly.

16. The Government of the Philippines shall take all necessary constitutional process for the implementation of the entire Agreement.

Fourth: This Agreement shall come into force with effect from the date of its signature.

Done in the City of Tripoli on 2nd Muharram 1397 H. corresponding to 23rd December 1976 A.D. in three original copies in Arabic, English, French languages, all equal in legal power.

FOR THE GOVERNMENT OF THE REPUBLIC OF THE PHILIPPINES:

Hon. CARMELO Z. BARBERO
Undersecretary of National Defense for Civilian Relations

Dr. ALI ABDUSSALAM TREKI
Minister of State for Foreign Affairs,
Libyan Arab Republic and Chairman of the Negotiations

FOR THE MORO NATIONAL LIBERATION FRONT:

Mr. NUR MISUARI
Chairman of the Front

Dr. AMADOU KARIM GAYE
Secretary General of the Organization of the Islamic Conference

APPENDIX 2

The Manifesto of the Muslim Independence Movement

IN THE NAME OF ALLAH, MOST GRACIOUS, MOST MERCIFUL.

PREAMBLE

The Muslim inhabitants of Mindanao, Sulu, and Palawan, invoking the grace of the Almighty Allah, Most Gracious, Most Merciful, on Whom all praise is due and Whom all creation depends for sustenance, make manifestation to the whole world its desire to secede from the Republic of the Philippines, in order to establish an Islamic State that shall embody their ideals and aspirations, conserve and develop their patrimony, their Islamic heritage, under the blessings of Islamic Universal Brotherhood and the regime of law, justice and democracy, and the recognized principles of the law of nations, do promulgate and make known the declaration of its independence from the mother country, the Republic of the Philippines.

TERRITORIES

The Islamic State shall comprise the contiguous southern portion of the Philippine Archipelago inhabited by the Muslims, such as, Cotabato, Davao, Zamboanga and Zamboanga City, Basilan City, Lanao, Sulu, Palawan, and the adjoining areas or islands which are inhabited by the Muslims or being under their sphere of influence, including the maritime areas therein.

PEOPLES

The Muslim inhabitants of the Republic of the Philippines, numbering some four millions (4,000,000) with culture and history of their own are distinct from the affluent Christian majority, and for the reasons abovementioned, its integration into the Philippines body politics being impossible.

DECLARATION OF PRINCIPLES

1. That it is a recognized principle underlying the Charter of the United Nations and the Declaration of Human Rights of the rights of all people constituting the minority in a given state for self-determination;
2. That the Islamic World Congress has affirmed the above principles, particularly on the rights of the Muslims who are in the minority in non-Muslim states for self-determination;
3. That the systematic extermination of the Muslim youth—like the Corregidor Fiasco—the policy of isolation and dispersal of the Muslim communities have been pursued vigorously by the government to the detriment of the Muslims;
4. That Islam, being a communal religion—an ideology and a way of life, must have a definite territory for the exercise of its tenets and teachings, and for the observance of its laws;
5. That economic progress, social development, and political independence are the cherished and inviolable dreams and aspirations of the Muslims, and the realization of which can better be served and promoted by and among themselves; and
6. That under the present state of things, the Muslims are capable of self-government or political independence they being endowed with sufficient number of professionals with academic, technical and legal training.

Now, therefore, it was decided that a government shall be organized and established for the Muslims inhabiting the aforesaid particular areas of Islands of Mindanao, Sulu and Palawan, and the outlying territories which are all under the sway of the Muslims, said government to be independent from all states and be equal with all others under the laws of civilized nations; and

Be it further known that said government for the Muslims shall be known and referred to as—

THE REPUBLIC OF MINDANAO AND SULU.

Pagalungan darul Islam, Cotabato, Philippines, May 1, 1968.

PRINCIPAL SIGNATORY:

(Sgd.) EX-GOVERNOR DATU UDTOG MATALAM

APPENDIX 3

The Manifesto of the Moro National Liberation Front

ESTABLISHMENT OF THE BANGSA MORO REPUBLIK

We, the five million oppressed Bangsa Moro people, wishing to free ourselves from the terror, oppression and tyranny of Filipino colonialism which has caused us untold sufferings and miseries by criminally usurping our land, by threatening Islam through wholesale destruction and desecration of its places of worship and its Holy Book, and murdering our innocent brothers, sisters and folks in a genocidal campaign of terrifying magnitude;

Aspiring to have the sole prerogative of defending and chartering our own national destiny in accordance with our own free will in order to ensure our future and that of our children;

Having evolved an appropriate form of ideology with which the unity of our people has been firmly established and their national identity and character strengthened;

Having established the Moro National Liberation Front and its military arm, the Bangsa Moro Army, as our principal instrument for achieving our primary goals and objectives with the unanimous support of the great mass of our people; and finally

Being now in firm control of a great portion of our national homeland through successive and crushing victories of our Bangsa Moro Army in battle against the Armed Forces of the Philippines and the Marcos military dictatorship, hereby declare:

1. That henceforth the Bangsa Moro people and Revolution, having established their Bangsa Moro Republik, are throwing off all their political, economic and other bonds with the oppressive government of the Philippines under the dictatorial regime of President Ferdinand E. Marcos to secure a free and independent state for the Bangsa Moro people;

2. That we believe armed struggle is the only means by which we can achieve the complete freedom and independence of our people, since Marcos and his government will never dismantle the edifice of Philippine colonial rule in our national homeland of their own accord;

3. That the Moro National Liberation Front and its military arm, the Bangsa Moro Army, shall not agree to any form of settlement or accord short of achieving total freedom and independence for our oppressed Bangsa Moro people;

4. That the Revolution of the Bangsa Moro people is a revolution with a social conscience. As such it is committed to the principle of establishing a democratic system of government which shall never allow or tolerate any form of exploitation and oppression of any human being by another or of one nation by another;

5. That those Filipinos who may wish to remain in the Bangsa Moro national homeland even after independence shall be welcomed and entitled to equal rights and protection with all other citizens of the Bangsa Moro Republik, provided that they formally renounce their Filipino citizenship and wholeheartedly accept Bangsa Moro citizenship; their property rights shall be fully respected and the free exercise of their political, cultural and religious rights shall be guaranteed;

6. That the Bangsa Moro people and Revolution are committed to the preservation and growth of Islamic culture among our people, without prejudice

to the development and growth of other religious and indigenous cultures in our homeland;

7. That our people and Revolution recognize and adhere to the Charter of the United Nations and the Universal Declaration of Human Rights; and, in addition, they shall respect and adhere to all laws binding upon the nations of the world;

8. That the Bangsa Moro people and Revolution are committed to the preservation and enhancement of world peace through mutual cooperation among nations and common progress of the peoples of the world. Accordingly, they are committed to the principle of mutual respect and friendship among nations irrespective of their ideological and religious creed;

9. That our people and the Revolution, upholding the principle of self-determination, support the right of all peoples of all nations in their legitimate and just struggle for national survival, freedom and independence;

10. That the Bangsa Moro people and the Revolution shall, in the interest of truth, guarantee the freedom of the press;

11. That, in order to accelerate the economic progress of our war ravaged Bangsa Moro homeland, our people and Revolution shall encourage foreign investment under terms and conditions beneficial to our people and the investors. Accordingly, those foreign investors in the Bangsa Moro homeland who may decide to continue their economic activities under the revolutionary regime shall be welcomed;

12. The Bangsa Moro people and the Revolution are committed to the principles that they are a part of the Islamic World as well as of the Third World and of the oppressed colonised humanity everywhere in the world.

Therefore, we hereby appeal to the conscience of all men everywhere and the sympathy of all the nations of the world to help accelerate the pace of our people's Revolution by formally and unequivocably recognizing and supporting our people's legitimate right to obtain their national freedom and independence. Such recognition and support must be concretised by accepting the Bangsa Moro Republik as one of the members of the family of independent and sovereign nations in the world and giving official recognition to the Moro National Liberation Front.

Done in the Bangsa Moro Homeland, this 28th day of April 1974.
Hajji Nur Misuari,
Chairman,
Central Committee,
Moro National Liberation Front.

APPENDIX 4

Structure of the Moro Organizations

It should be noted at the outset that all the organizational structures described below are constructed on the basis of organization charts depicted on paper by the respective fronts. Figures A4.1–A4.4 show the basic structure of the MNLF-Nur Misuari Faction as revealed by Hatimil Hassan, Director of MNLF Foreign Relations Office in Syria, and by Commander Solitario, Chairman of Ranao Norte Revolutionary Committee (interview, Commander Solitario, Lanao del Norte, Philippines, 17–20 February 1985). According to Hatimil Hassan (1981: 253–6), the MNLF is a democratic people's organization which functions on a committee system. It is governed by a Central Committee headed by a Chairman who in theory is elected by a group of delegates chosen at the provincial level. The Central Committee members are selected and appointed by the Chairman. Since its official formation in 1974, the membership of the Central Committee has undergone constant changes due to individuals surrendering, defection to other factions, and resignation. The Central Committee has a secretariat that serves as the executive and administrative body of the organization; it consists of different bureaux or committees which perform various functions indicated in Figure A4.1. Each bureau is headed by a chairman or a director who is an ex-officio member of the Central Committee.

The MNLF comprises a Congress and a Revolutionary Tribunal which serve as legislative and judicial bodies respectively. In practice, the National Congress is held irregularly, when consultations are needed to ratify or make important decisions. The MNLF has held only two national congresses since its formation. The first National Congress was convened in Zamboanga in 1974 to ratify the Front's manifesto. The second was held on the island of Jolo on 2–6 September 1986 on the occasion of a historic meeting between the President of the Philippines, Mrs Corazon Aquino, and the Chairman of the MNLF, Mr Nur Misuari.

Figure A4.1
Structure of the MNLF Organization (A)

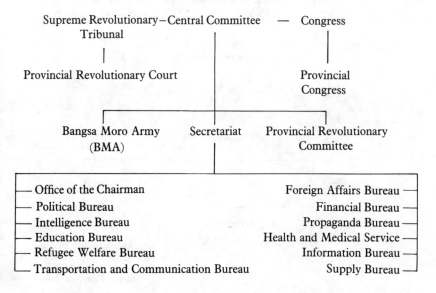

Figure A4.2
Structure of the MNLF Organization (B)

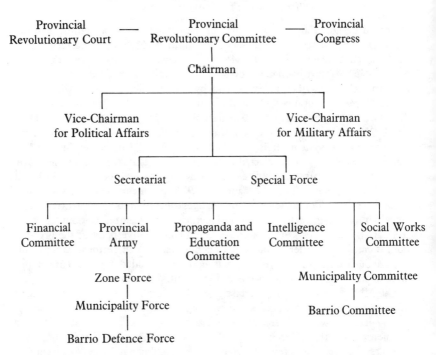

The Provincial Congresses, which are held by the Provincial Revolutionary Committees, have convened from time to time. The Revolutionary Tribunal, consisting of the Supreme Revolutionary Tribunal and Provincial and Municipal Courts, assumes primarily civil functions. The Military Tribunal is created on an *ad hoc* basis by the General Staff of the Bangsa Moro Army (BMA), the military force of the MNLF. In 1986, it is estimated that the MNLF had about 10,000 regular BMA members, of which about 50 per cent were heavily armed

Figure A4.3
Structure of the MNLF Organization (C)

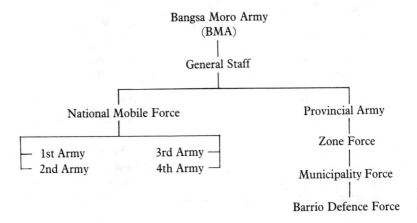

Figure A4.4
Structure of the MNLF Organization (D)

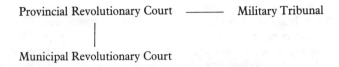

(*Asiaweek*, 14 September 1986). They are orgnized into ten provincial armies and a number of zones in each army (Hatimil Hassan, 1981: 255–6). The BMA also provides the National Mobile Forces, divided into four armies, which occupy at least 13 permanent camps throughout the region (*Far Eastern Economic Review*, 11 September 1986). In addition, some Moro villagers are armed and function as *barrio* defence forces.

The structural organization of the MNLF may be regarded as having a 'secular-orientation'. Another slightly different structural emphasis is the 'Islamic-oriented' one. This is not to suggest that the former is less 'Islamic' than the latter, for Islam does not restrict a Muslim from doing what he pleases as long as he does not contradict Islamic teachings. The MNLF-Reformist Faction employs the former, while the BMLO (BMILO) and the MILF use the latter. Since the MNLF-Reformist Faction still uses basically the same structural pattern as that of the Misuari Faction, except for those reforms made in accordance with the Nine-point Proposal (Appendix 5), is not necessary to elaborate its structural organization.

The BMLO was first to emphasize the 'Islamic-oriented' organization. In its policy statement issued in May 1978, the BMLO outlined the structure shown in Figure A4.5 (BMLO, 1978: 2–3).

Like the MNLF, the BMLO is governed by a Supreme Executive Council (SEC) headed by a Chairman. Unlike the MNLF, the choice of a chairman of the BMLO is agreed upon by its members through *shura* (Islamic-oriented meeting). Membership of the SEC is based on merit and on regional and ethnic representation. The SEC operates through different committees of the secretariat, which functions as the executive and administrative body. The Consultative Assembly of the Moro People (CAMP) is the highest policy-making body of the BMLO. All members of the SEC, and the high-ranking members of other or-

Figure A4.5
Structure of the BMLO Organization

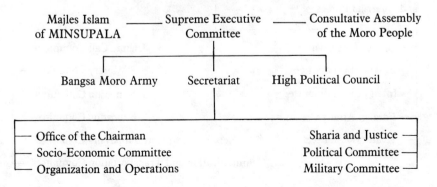

gans of the BMLO, are provisional members of CAMP. The CAMP holds a congress yearly to approve policies and budgets for each fiscal year. Special meetings of the CAMP may be held when needed, but not more than once per month. The significant organ that makes the BMLO an 'Islamic-oriented' front is the Majles Islam of MINSUPALA. It consists of *ulama* headed by a *mufti*. The Majles Islam of MINSUPALA acts as Supreme Judicial Tribunal and has power to repeal or amend any orders that do not conform to Islam. It is also responsible for the enforcement of the *Sharia* on members of the BMLO and on Muslims under its jurisdiction. In fact, the Mufti is considered one of the most powerful leaders of the BMLO. The Bangsa Moro Army (BMA) is the *kampilan* (sword) of the BMLO. The *mujahideen* (fighters) of the BMA fight only in the name of Islam and for the Moro homeland. Able Muslims who are willing to join the struggle are trained for membership in the BMA. The High Political Council (HPC) consists of leaders of different élite groups, such as sultans, *datu*, *ulama*, and professionals who are committed to the Islamic struggle of the Moro people. The HPC, with the approval of the SEC, conducts the political affairs and formulates the ideology of the BMLO (BMLO, 1978).

Figure A4.6
Structure of the MILF Organization (A)

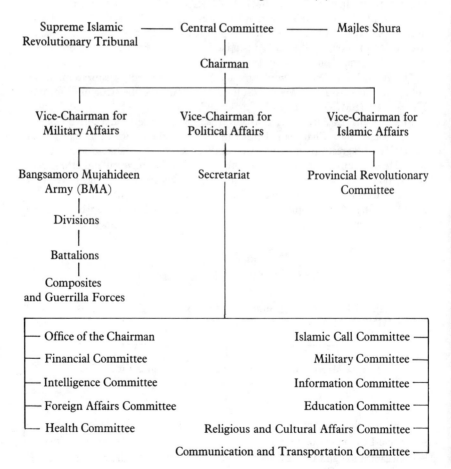

Figure A4.7
Structure of the MILF Organization (B)

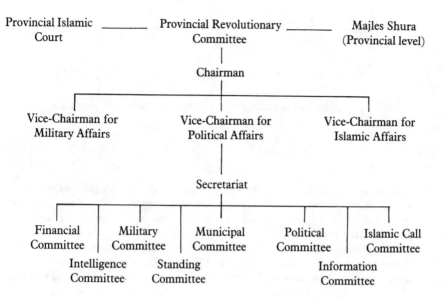

During the author's visit to the MILF camps (Camp Busrah in Lanao del Sur, Camp Abubakar in North Cotabato, and Camp Omar in Maguindanao), certain field commanders constructed the MILF organization, as illustrated in Figures A4.6–A4.7.

Like the BMLO, the MILF is an Islamic-oriented organization. The Supreme Islamic Revolutionary Tribunal and the Majles Shura of the MILF, corresponding to the Majles Islam of MINSUPALA and the Consultative Assembly of the Moro People of the BMLO, respectively, are organs that differentiate the Islamic from secular-oriented fronts. That is, the MILF and the BMLO use Islamic court and *shura*, whereas the MNLF employs civil or military courts and its congress functions like a Western organization. Furthermore, the MILF and BMLO consider themselves as vehicles of *jihad* for Muslims in Mindanao (Hashim, 1985: 24; BMILO, 1984a: 1). They regard *jihad* as a collective duty for which every Muslim has a responsibility—an obligation that accepts no exception. To stress the necessity of organization in waging a *jihad*, Hashim (1985: 25) writes, 'The wisdom and lesson we can imbibe from the Prophet's experience is clear—that to fight alone is not Jihad, to fight in a disorganized fashion is neither [*sic*] Jihad.' More importantly, the Islamic-oriented fronts categorically reject the concept that *jihad* can be fought under a non-Islamic group or state (Hashim, 1985). This is the underlying reason why the MILF and BMLO advocate an Islamic line.

The fifth group is the Moro Revolutionary Organization (MORO). 'Mike', one of the MORO's leaders, indicates that the group adheres to the structural pattern of the CPP. It includes the basic units as demonstrated in Figure A4.8.

According to the constitution of the CPP, the Central Committee is the highest authority of the organization when the Congress is not in session. However, when it is, the Congress dictates the overall work of the organization and defines its programmes, including the election procedure of the Central

Figure A4.8
Structure of the MORO Organization

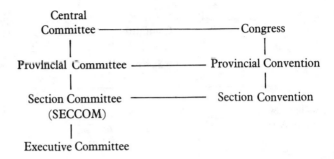

Committee. All members of the Central Committee are delegates to the Congress, which convenes yearly. The Provincial Committee, whose members are elected during the Provincial Convention, functions as the central committee at the provincial level, and the Provincial Convention as that of the Congress. The Section Committee (SECCOM) comprises the nuclei organized within one specified section of the city or town. The SECCOM Convention, which meets once a year, makes final decisions concerning the work and plans of the SECCOM. The lowest organ of the organization is the Executive Committee, which runs the nucleus established in particular places (CPP, 1969: 74–81).

APPENDIX 5

The Nine-point Proposal of the Reformist Group

'BISMILLAAHIR RAHMAANIR RAHIIM'

MORO NATIONAL LIBERATION FRONT
General Headquarters
(REFORMIST)

OFFICE OF THE CHAIRMAN

7 March 1982

Brothers et al.:

Assalamu alaikum wa rahmatullahi wa barakatuh!

In this letter, please find out the Nine (9) Point proposals that the Reformist Group submitted to Nur Misuari for consideration for the progress and liberation of our oppressed people from the clutches of oppression, suppression, exploitation and other ills of the Bangsa Moro Revolution, To Wit:

1. Real, functional and representative Central Committee. This means that the Central Committee of the MNLF shall be represented by every tribe of the province of origin. It must be working and functional as a body in laying down the National Policies, Principles, Guidelines and Programs of the Islamic Revolution.

2. Creation of the Executive Committee to implement the policies and programs of the Central Committee on the ground. This Committee shall be responsible to administer [sic] and supervise the policies and programs of the MNLF. It shall compose the responsible leaders of the Islamic Revolution.

3. Creation and/or implementation of the Steering Committee. This Committee shall be composed of the Provincial Chairmen, Vice-Chairmen, Provincial Commanders and mass leaders in the promulgation of the policies and programs of the province.

4. Strengthening and revitalization of all the branches and organs at the national level. It means that all national organs shall be defined and manned by dedicated and qualified personnel/members of the MNLF.

5. Strengthening of the Foreign Committee shall be encouraged to accommodate more qualified personnel from different provinces to render services abroad.

6. Implementation of the Central Committee Resolution in the Tripoli Meeting to last letter.

7. Sincere efforts in forging unity with other organizations. It means that all efforts shall be exhausted in seeking unity with the BMLO, MNLF Salamat Faction and other organizations interested to the liberation of our people, homeland and Islam.

8. Localization of leaders in all levels and commands in the province. It means that the natives of every province shall have the first priority to handle the affairs of the province.

9. Promulgation/implementation of the Constitution and By-Laws of the Islamic Revolution. In order to run the Revolution, there must be a constitution and by-laws for the direction and proper course of the Islamic Revolution. This is to safeguard the interest of the Islamic Revolution.

(Signed)
The Reformist Group
in the 13 Provinces

APPENDIX 6

Resolutions of the Reformist Group:
Rejection of Misuari's Leadership and Acceptance of Autonomy

MORO NATIONAL LIBERATION FRONT

National People's Congress
3rd Session

RESOLUTION NO. A-2
CONFIRMATION OF THE OUSTER OF NUR MISUARI
AS CHAIRMAN OF THE CENTRAL COMMITTEE

WHEREAS Nur Misuari's betrayal to the cause of the Bangsa Moro people homeland and Islam caused untold sufferings and sacrifices to millions of the Bangsa Moro people and spilt the blood of the Mujahideen infavor [sic] of his personal glory, selfish motives and evil designs;

WHEREAS the leftist leanings of Misuari which derailed the direction of the Revolution towards its real objective of defending the cause of Islam is not acceptable to the great masses of our people and repugnant to the senses of the Mujahideen;

WHEREAS the refusal of Misuari to recognize the Tripoli Agreement preventing it from being implemented to the prejudice of the Bangsa Moro people and in defiance of the Organization of the Islamic Conference Resolutions and the advice of the friendly countries caused the unpopularity of the MNLF to [sic] such bodies;

WHEREAS his failure to unify and accommodate all sectors of the Moro society, the professionals, farmers, workers, students, intellectuals, politicians and traditional leaders prevented the clinching of the Revolutionary victory and have dragged the Revolution to factionalism;

WHEREAS his failure to consult the leaders on the Ground on major decisions such as the Tripoli Agreement, reverting it to Independence, his maladministration, corruption, one-man rule, and other malpractices destroyed the morale of the Fighting Forces which sapped the trust and confidence of its leaders and commanders on the Ground;

WHEREAS the presentative members of the Central Committee have decided and did oust Misuari from the present position in the proper Resolution on the same;

THEREFORE IT IS HEREBY RESOLVED that the National People's Congress hereby confirmed the OUSTER of Misuari from the Chairmanship of the Moro National Liberation Front Central Committee and all the powers, functions, privileges and rights inherent or attendant thereto is hereby stripped and he be further advised to cease and desist to function as such;

RESOLVED FINALLY that this Resolution be submitted to the Organization of the Islamic Conference and its member States be furnished a copy.

RESOLUTION NO. A-6
ACCEPTANCE OF THE AUTONOMOUS STATE

WHEREAS the Tripoli Agreement forged between the Moro National Liberation Front and the Philippine Government signed on December 23, 1976 creating the autonomous State within the sovereignty and the territorial integrity of the Republic of the Philippines;

WHEREAS both Signatories to the said Agreement have in one way or the other become apathetic and evaded to negotiate to its final stage thereby preventing the implementation of the Agreement in its real form and purpose;

WHEREAS the Bangsa Moro people are suffering and shouldering the brunt of the war as a result of the genocidal campaign waged by the colonial forces of the Marcos regime and the irresponsible reaction of the parties resulted to [sic] the non-implementation and the violation of the Tripoli Agreement;

WHEREAS it is the wishes of the great majority of the Bangsa Moro people to accept the Autonomous State as envisioned under the Tripoli Agreement through a negotiation under the auspices of the Organization of the Islamic Conference to be represented by their real leaders contrary to the stand of Misuari in reverting his stand to independence;

WHEREAS it is the position of the majority of the leadership of the Bangsa Moro Revolution including the real Mujahideen and their commanders in the Field that the Tripoli Agreement is an international covenant that must be respected and recognized by the real parties;

NOW THEREFORE, FOR AND IN CONSIDERATION OF THE ABOVE, IT IS HEREBY RESOLVED that the Organization of the Islamic Conference be informed of the wishes of the majority of the Leadership of the Revolution and the great majority of our people to accept the Autonomous State envisioned under the Tripoli Agreement.

RESOLVED FURTHER that the Organization of the Islamic Conference be requested to continue its role in seeking ways to bring the real parties to the negotiating table in order to put an end to the conflict in South Philippines.

RESOLVED FURTHERMORE that the Organization of the Islamic Conference be reminded of the change of leadership in the Moro National Liberation Front so that the proper parties be represented by the proper representatives to finally resolve the issue.

RESOLVED LASTLY that copies of this Resolution be furnished to the Organization of the Islamic Conference and its member states and the Philippine Government for proper information.

IN ATTESTATION THEREOF, we hereunto affixed our signature this 10th of June 1982 in mainland Mindanao.

Signature

(Forty-three leaders and members of the MNLF-Reformist Group)

I HEREBY CERTIFY THAT THE FOREGOING IS TRUE AND CORRECT. This Resolution had been deliberated on the floor during the Third Session of the Moro Congress from [sic] June 8, 9 and 10, 1982.

> KIRAM HADJI KAMALUDDIN
> Secretary-General
> Member, Moro Congress
> Tawi-Tawi

APPENDIX 7

The MORO Manifesto

We, the *Moro Revolutionary Organization* (MORO), hereby declare our basic principles and general program.

1. *Unity of the Moro People for National Self-Determination and Democracy*

The broad masses of the Moro people must be united in the revolutionary struggle for national self-determination and democracy against U.S. imperialism, feudalism and bureaucrat capitalism. Our main enemy now is the fascist dictatorial regime of the U.S.-Marcos clique which is the concentrated expression of the three aforementioned evils.

The Moro people include the ethno-linguistic groups of the Maranaos, Maguindanaos, Tausugs, Samals, Yakans, Iranuns, Kalagans, Badjaos, Palawanis, Jama Mapuns, Kalebugans, Sangils and other such people who have intimate historical bonds with the foregoing and are willing to be integrated with the Moro people in any area (province, municipality, city or barrio) where the Moro people are the majority nationality.

Since the Spanish colonial times, the foreign rulers and domestic ruling classes of the Philippines have always sought to subject the Moro people to national oppression and exploitation. The Moro people have always put up a heroic resistance in accordance with their just interests.

In the period of its direct colonial rule and in the current period of its indirect semicolonial rule over the Philippines, U.S. imperialism has carried out a policy of subjecting the Moro people to national oppression and exploitation. Not only the domestic ruling classes of the Philippines but also of the Moro areas have been the willing tools of U.S. imperialism.

National exploitation is the reason behind national oppression and the chauvinism. U.S. imperialism and the Filipino and the Moro ruling classes are selfishly interested in the exploitation of the broad masses of the Moro and Filipino people and in the plunder of the rich natural resources of the Minsupala region.

In upholding the right to national self-determination, the Moro people can go to the extent of fighting for secession from the present reactionary semicolonial and semifeudal Philippine state. They can opt for regional autonomy only in a people's democratic state which can guarantee the quality of nations in the Philippine archipelago.

The development of the Philippine revolution cannot but be uneven. The Moro people can go at their own rate in seeking to establish a people's democratic state. If they can, they can go ahead of the rest of the Filipino people in establishing such a state. At the same time, they should always be willing to unite, cooperate and coordinate with the Filipino people in fighting for common interests and aspirations against the same enemies.

The struggle for national self-determination must be linked to the struggle for democracy. Among the broad masses of the Moro people, especially among the working class and the peasantry, there is no point in struggling for national self-determination if this is not substantiated by a democratic revolution.

Otherwise, the domestic ruling classes in the Moro areas will continue to be in a position to make deals with the enemy in the manner of the Bates Treaty and sell out the interests of the people. Colonial domination is not possible without the collaboration of the capitulationists, renegades and traitors.

A democratic revolution must serve as the main content of the struggle for national self-determination. This does not simply mean the restoration of formal bourgeois democratic rights and procedures. Essentially, it means the solution of the land problem of the peasant majority of the Moro and non-Moro people in the Moro areas.

This problem has been aggravated by wanton landgrabbing, by the foreign monopoly capitalists, and by such local exploiters as the big compradors, the big landlords, and the big bureaucrats of the reactionary state. Plantations, logging, mines and industrial estates have been established on Moro land to exploit the Moro and non-Moro people.

The big traitors to the revolutionary cause of the Moro people also participate in grabbing land, in titling land to themselves economically and are thus willing to become running dogs of the Manila government and the foreign monopoly capitalists.

The seas of the Moro people are also being encroached upon by foreign petroleum mining companies and fishing companies. Thus, the foreign monopoly capitalists and their local stooges are extremely antagonistic to the Moro people's revolutionary cause.

A national democratic revolution must strike at the roots of national oppression and exploitation. It is the sacred duty of the *Moro Revolutionary Organization* to arouse, organize and mobilize the broad masses of the Moro people along the general line of the national democratic revolution.

The toiling masses of workers and peasants (including the fishermen and farmworkers) must first of all be united to serve as the basic solid foundation of the great national unity and revolutionary struggle of the Moro people against their oppressors and exploiters.

The petty bourgeoisie (the self-sufficient owners of some property and the intelligentsia including the students, teachers and professionals) are also an important part of the Moro people's revolutionary unity and struggle. So are the national bourgeoisie or the middle bourgeoisie which should be attracted to the worker–peasant–petty bourgeois revolutionary alliance.

The lower and enlightened sections or elements of the exploiting classes may be allowed a positive role in the struggle. Moreover, the split among the reactionaries can be taken advantage of by the revolutionaries. It is possible to have unstable, temporary and indirect allies in the struggle.

The revolutionary struggle of the Moro people must be comprehensively national democratic in character. It is anti-fascist, anti-imperialist and anti-feudal. At the vanguard of the revolutionary movement of the Moro people, it should be the Moro cadres who grasp the ideas of the most advanced revolutionary class and implement them.

The perspective of this movement is socialist. Therefore, it is absolutely necessary that the revolutionary proletariat is the leading class in the Moro revolution not only for the purpose of winning victory in the present national democratic stage but also consequently for moving on to the socialist stage of the revolution.

We, the *Moro Revolutionary Organization*, are dedicated to promoting the general line of the national democratic revolution among the Moro people and developing the core of proletarian revolutionary cadres among the Moro people. Militant revolutionary political and organizational work must be done for the purpose.

The *Moro Revolutionary Organization* must build and strengthen itself ideologically, politically and organizationally. It should create and strengthen mass organizations, people's armed defense units and organs of democratic power so as to carry forward the revolutionary struggle of the Moro people. MORO groups are to be formed within these to serve as the revolutionary hard core.

Our association is interested in developing unity, cooperation and coordination with the Moro National Liberation Front, the Bangsa Moro Army and all other organizations and individuals which are interested in advancing the revolutionary cause of the Moro people.

2. *People's War as the Main Form of the Moro People's Revolutionary Struggle*

People's war, wherein the revolutionary political struggle is coordinated with the revolutionary armed struggle, is the main form of the revolutionary struggle of the Moro people for national self-determination and democracy. The Moro people must consistently wage revolutionary armed struggle for so long as they are subjected to national oppression and exploitation by U.S. imperialism, feudalism and bureaucrat capitalism.

At the same time, various forms of revolutionary political struggle must be employed, developed and put into coordination with the revolutionary armed struggle. By using all possible forms of struggle, the people can undermine the anti-Moro and anti-democratic state. In the final analysis, however, it is the people's armed power that can overthrow the rule of this state over the Moro people.

Because their revolutionary cause is just, the Moro people are always ready to fight until total victory is won. Otherwise, they can only be subjected to the peace of slaves imposed by the perpetrators of national oppression—the fascist, the big foreign and domestic capitalists and the big landlords.

So many lives of the Moro people have been sacrificed and so much blood of the Moro people has been shed in brutal massacres and murder campaigns waged by the present reactionary state of the Philippines. In countless acts of pillage and plunder, the enemy has destroyed the homes and livelihood of the people, created multitudes of war refugees, subjected so many women to rape and rounded up so many people on whom torture and other indignities are applied.

It is a grievous sin not to avenge these crimes and render justice to the victims. Without asserting and fighting for revolutionary justice, the Moro people will continue to be subjected to a daily life of national oppression and exploitation, no less brutal and even worse than the massacres used to impose the unjust rule of the enemy.

The heroic Moro people are ready to wage revolutionary armed struggle no matter how long this shall take. It is because they fight for national self-determination and democracy. They can rely on themselves. They have more than enough arms to start with and [sic] pursue a self-reliant struggle.

The people's war cannot but be protracted. It takes time to change the balance of forces between us and the enemy. The enemy starts with a military strength superior to ours. On our part, we seek to strengthen ourselves by defeating his forces piece by piece and seizing more and more arms from them.

Our people's war may take three strategic stages. The first stage is that we are on the defensive while the enemy is on the offensive. The second stage is that we have achieved a stalemate with the enemy. The third and final stage is that we are the ones on the offensive while the enemy is on the defensive. We shall certainly reach this final stage, the stage of completely liberating ourselves and completely ridding the Moro areas of the hated reactionary armed forces of the Philippines.

Even as we are on the strategic defensive now, we can and must launch tactical offensives. Citing the heroic feats of many armed units of the Bangsa Moro Army and the New People's Army, we can see that it is possible to launch tactical offensives in the form of guerrilla warfare and at certain times in the form of regular mobile warfare.

With the people's support, the revolutionary armed forces can force the enemy to spread out his forces thinly over wide areas. And in launching tactical offensives, we raise a large armed force superior to that enemy unit which we choose to strike at. Thus, we can defeat and wipe out enemy units repeatedly, accumulate strength after every battle and change the balance of forces step by step.

Our style of fighting should be characterized by surprise and mobility and flexibility of tactics. We should fight only battles that we are sure of winning. These are battles where we have the initiative, where our strength is superior to the enemy's and where we can annihilate units of the enemy and seize arms.

The tactics of guerrilla warfare include: dispersal of the guerrilla units for the purpose of developing a wide mass base and consequently forcing the enemy to spread his forces, concentration of forces for the purpose of launching tactical offensives against the enemy; and shifting for the purpose of avoiding encirclement by a superior enemy force.

Guerrilla forces or regular mobile forces should lure enemy units deep into the rural areas where they have the political and military advantages or where they can gain the upperhand. At the present stage, it is best to maintain mobility to be able to destroy one enemy unit after another.

At the present level of armed strength of the Bangsa Moro Army, its fighting units can besiege or prick one enemy post to destroy enemy reinforcements at an unexpected point along the way or strike at another enemy post which can be destroyed easily as a result. The use of feints can lead the enemy to commit mistakes and enhance the initiative of the revolutionary armed forces.

People's war is not only a question of military strategy and tactics. It is also a question of laying down the political foundation for the success of the military strategy and tactics. The best military commander could be defeated if his unit is lacking in support from the people. His unit could be easily isolated and turn [sic] into a bandit unit if it has no popular support and no correct political direction.

In a people's war that is being carried out in the Moro areas, it is fundamental to arouse, organize and mobilize the masses, carry out land reform step by step and exert efforts to transform the backward villages into politically, economically, militarily and culturally advanced villages.

The people's army can get the support of the people only if the people are convinced that this army is working and fighting for their interests. The people themselves are the endless source of military personnel, intelligence, logistics and other requirements of revolutionary war.

We, the *Moro Revolutionary Organization*, are determined to create and develop people's armed defense units. We are willing to coordinate these units with other units devoted to the revolutionary cause of the Moro people. We are also willing to help develop those armed units established outside of our initiative in the Moro areas.

In the Moro areas, the armed units of non-Moro people should unite, cooperate and coordinate with the armed units of the Moro people. We should also encourage the development of the people's armed units in which Moro and non-Moro fighters are integrated. In this way, we can defeat the enemy's policy of dividing and ruling the people of the Moro areas.

In the establishment of mass organizations, local armed units and organs of democratic power, our policy is likewise to encourage the integration of the Moro and non-Moro people so as to give the broadest possible support to the people's armed units and so as to defeat the enemy's policy of making the Moro and non-Moro people suspect and fight each other.

The *Moro Revolutionary Organization* is wholeheartedly devoted to the revolutionary cause of the Moro people. An integral part of this devotion is to help integrate well the non-Moro people with the Moro people in Moro areas and elsewhere.

3. *Unity of the Moro People with the non-Moro People in the Moro Areas, Minsupala and in the Philippines*

The Moro people must be united with the non-Moro people in the Moro areas, in the whole Minsupala and in the entire Philippine archipelago. On the three aforesaid scales, the Moro and the non-Moro people have a common interest in the national democratic revolution and are faced in common with the same enemies—U.S. imperialism, feudalism and bureaucratic capitalism.

The Moro people are in the overwhelming majority in Sulu, Tawi-Tawi, Basilan, Lanao del Sur and Maguindanao and many municipalities and barrios in North Cotabato, Maguindanao, Sultan Kudarat, Lanao del Norte, Zamboanga del Norte, Zamboanga del Sur, the Davao provinces, Bukidnon, Palawan and Cagayan de Sulu.

But in many of these areas, there are considerable numbers of people belonging to other nationalities, including settlers and native inhabitants who are also non-Moro. Also adjacent to the areas where the Moro people prevail in number are greater masses of non-Moro people whose lives cannot but be related to those of the Moro people.

It is to the interest of the Moro people to unite and cooperate with the non-Moro people in the Moro areas and at the same time give due respect to their national peculiarities. At the base of the broad unity and cooperation to be fostered should be the unity and cooperation of the oppressed and exploited classes of workers and peasants.

The Moro and the non-Moro people should participate in the same mass organizations, armed units and organs of democratic power and in the activities thereof. The rule of proportionate representation in the organs of leadership should be applied as far as possible as a safeguard against chauvinism and discrimination.

Membership in our association, the *Moro Revolutionary Organization*, is open not only to Moros but also to non-Moros who adhere to the basic principles and general program as laid down in this Moro Manifesto and who pledge to work and fight for the revolutionary interests of the Moro people.

The same rules and procedures for enhancing harmony between the Moro and non-Moro people should be followed in the areas of Minsupala where the Moro people are in the minority and the non-Moro people are in the majority. Always, the overriding guide should be to advance the revolutionary struggle along the national democratic line.

Of the total population of Minsupala, the Moro people comprise about one-third and the non-Moro people, two-thirds. It is absolutely necessary for the Moro and non-Moro people in Minsupala to unite, cooperate and coordinate their revolutionary efforts and frustrate all schemes to divide and rule them.

It is adverse to the interest of the Moro people to allow or promote any sectarianism or narrowmindedness that would tend to divide them from the non-Moro people. Only the enemy and his instrumentalities are interested in fomenting chauvinism, sectarianism and communal conflicts so as to prevent the people, especially the exploited and oppressed masses, from uniting against him and for national democratic revolution.

As far as the entire Philippine archipelago is concerned, it is a clear objective fact that the revolutionary struggles of the Moro people and the Filipino people support each other. We also learn from history that Spanish colonialism collapsed and the Japanese fascist invasion failed as soon as the common armed resistance of the Moro and non-Moro people in the Philippines reached a high level of development.

It is politically wise and necessary to enhance the unity, cooperation and coordination between the revolutionary forces of the Moro people, including the Moro National Liberation Front and the Bangsa Moro Army, on the one hand, and those of the Filipino people, including the Communist Party of the Philippines and the New People's Army, on the other. To do so is to strengthen all revolutionary forces in the Philippines and hasten the downfall not only of the fascist dictatorial regime of the U.S.–Marcos clique but also of the entire reactionary state oppressing and exploiting the people.

So long as the reactionary state and ruling class can dominate the greater part of the archipelago, they are in a position to prolong their rule over the Moro people. It is to the interest of the Moro people to extend support to the Filipino people, just as it is to the interest of the latter to do likewise to the former.

Because of the continued and rising revolutionary struggles of the Filipino and Moro people, the enemy is forced to divide his forces and spread them thinly all over the archipelago. His ability to concentrate the reactionary armed forces anywhere is lessened. In the years to come, the enemy will find himself more and more overextended, more and more exhausted and more and more vulnerable to the deadly blows of revolutionary forces throughout the archipelago.

In the Moro areas, the Moro people have long given due respect and encouragement to various forms of organizations and institutions initiated by non-Moro people. We are glad to acknowledge that in all non-Moro areas, where there are Moro communities, students and others, the Filipino people accord due respect and encouragement to Moro organizations and institutions.

It is only the foreign and domestic oppressors and exploiters of the Filipino people and the Moro people who are interested in national oppression, chauvinism and discrimination.

We, the *Moro Revolutionary Organization*, are committed to promoting in every possible way the common understanding and common courses of action between the Moro people and other people in the Philippine archipelago so that revolution can triumph along the national democratic line.

4. *All-Round Progress and Religious Freedom*

The broad masses of the Moro people demand all-round progress. They seek progress in the political, socio-economic and cultural fields. They are opposed to the misrepresentation of the enemy that they could be appeased if only tokens of respect were given to the "cultural autonomy" and religion.

As they engage in the revolutionary struggle for national self-determination and democracy, they can achieve all-round progress to an extent as would enable them to defeat the enemy. The national democratic revolution which they must complete entails political, socio-economic and cultural requirements, aside from military.

By fulfilling all requirements of the revolutionary struggle in order to win victory, they strengthen themselves and lay the foundation for greater progress in all fields. The competition of the national democratic revolution means the beginning of socialist revolution and construction.

In the course of the national democratic revolution, the leadership of the most advanced class is realized and the broad masses of the people, mainly the workers and peasants, liberate themselves from national and class oppression and exploitation. Not just in the form but more importantly in substance, democracy is achieved and enjoyed by the people.

It means a [sic] gigantic progress to put a stop to national oppression and the unbridled plunder of human and natural resources by the foreign monopoly capitalist and the local exploiting classes. Only in the course of the revolutionary armed resistance, liberated areas arise and restraints as well as taxation can be increasingly imposed on the exploiters. Thus, the Moro people and their revolutionary government increasingly get a share of the social wealth.

The workers and masses can demand better working and living conditions. And in the countryside, where the people's army is based, land reform is undertaken by peasant masses and their associations as a step towards a throughgoing agrarian revolution. The backward villages can be transformed step by step into advanced bulwarks of the revolution.

The petty bourgeoisie also widen their opportunities against national discrimination, oppression and exploitation. The legitimate interests of the national bourgeoisie and the enlightened sections of the exploiting classes are protected.

In line with the political and socio-economic progress that they demand, the broad masses of the Moro people are profoundly interested in the development of a national, scientific and mass culture.

The past, including the revolutionary tradition and cultural heritage of the Moro people, is used to serve present-day revolutionary needs. And foreign things, especially in the scientific achievements of mankind, are used to serve the Moro people and land. The main consideration in any kind of progress is the progress that redounds to the toiling masses of workers and peasants.

We recognize that Islam has been a major factor in the historical development of the Moro people as a nation, especially in resisting the aggression of Spanish colonialism and of American imperialism and in the current struggle for self-determination. At the same time, we also recognize the democratic requirements fostered by contemporary realities and ideologies in the era of modern imperialism.

U.S. imperialism and the Philippine ruling classes have been able to lay effective control on the Moro people and land for several decades already. We must recognize our historical and national bonds with the broad masses of the Filipino people and the ever pressing necessity to unite, cooperate and coordinate with them in opposing U.S. imperialism and the ruling classes.

Cultural autonomism and religious sectarianism are tools of reaction. U.S. imperialism and the local reactionaries have used them to lull one people to cultural autonomism and religious sectarianism so that they can continue to divide and rule the people, preoccupy them with communal conflicts and give U.S. imperialism and the local reactionaries plenty of ground for manoeuvre against the entire people.

Notwithstanding all sectarian schemes of the chauvinists, both the Moro and non-Moro people in the Moro areas and the whole of Mindanao have shown the utmost tolerance for the diversity of faiths and have been averse to any religious persecution. There is a greater need and more progressive basis for the freedom of religious belief than religious sectarianism.

By and large, the fascist dictatorial regime of the U.S.–Marcos clique has failed to arouse the Filipino people against the Moro people on the basis of religious sectarianism despite the most desperate attempts to do so. It is necessary for the Moro and Filipino people to continue defeating the fascist regime and any succeeding reactionary regime in this regard.

We are for a secular democratic type of people's government, whether for the Moro areas or for the entire Philippine archipelago. It is a people's coalition government representing the interests of both the Moro people and the Filipino people. It is a government that guarantees the enjoyment of all democratic rights by the people. Among these rights is the freedom of religious belief.

Even before Philippine-wide or Moro-wide victory, when a significant part of the Philippines and the Moro areas shall have been liberated from enemy rule, the MORO can join hands with other revolutionary groups in establishing a provisional revolutionary coalition government, which can include even those serving under the present government who choose to serve the people.

5. Revolutionary Internationalism Against Imperialism

The Moro people's struggle for national self-determination and democracy must be guided by the principle of revolutionary internationalism. Under this principle, the Moro people can grasp fully the new democratic character of their revolution and the relation of this revolution to other revolutions in the world.

It is revolutionary internationalism for the Moro people to recognize that they have a share in the struggle of mankind for emancipation from imperialism, colonialism and all reaction and also for them to recognize that they draw support and learn from other people's revolutionary struggles.

To know the international context of their revolutionary struggle, the Moro people can grasp comprehensively not only the character of this struggle but also its strategic and tactical requirements. In the era of imperialism, no people waging a revolution can dispense with correct and appropriate international relations.

The Moro people must deal with U.S. imperialism as presently their No. 1 enemy among the foreign exploiters of their land and the entire Philippines. It is after all the No. 1 military supporter of the Marcos fascist gang.

The Moro people, together with the entire Filipino people, must count themselves among the people of the Third World and oppose imperialism. The people of Asia, Africa and Latin America have a common revolutionary cause with the Moro and Filipino people. They must support each other. The Moro people must also develop revolutionary unity and cooperation with the people of the world. There should be no restrictions in extending as well as seeking support from the people of the world.

In Southeast Asia, the Moro people have one of the biggest armed forces still fighting for national and social liberation. They are duty-bound to fight U.S. imperialism and the local reactionaries effectively until victory is won. The people in the unliberated countries of Southeast Asia continue to wage revolutionary armed struggle. The peoples of Southeast Asia must support each other.

We are opposed to the schemes of U.S. imperialism and the reactionaries of Southeast Asia to use bilateral agreements between the Philippines and neighbouring countries and the Association of Southeast Asian Nations (ASEAN) against the revolutionary people. It is correct and appropriate for the Moro people to call on the other peoples of Southeast Asia to frustrate these schemes and resort to common efforts to frustrate them.

The Moro people have close bonds with the people of all Islamic countries and with the Arab and Palestinian peoples. We make special mention of our unity with them in seeking liberation from imperialism, colonialism and hegemonism and in achieving a world of justice, peace, progress and prosperity.

The Moro people support the revolutionary struggles of the Arab and Palestinian peoples against Israeli Zionism and the superpowers which give direct or indirect support to it. The national rights of the Palestinian people must be restored to them. And the territories seized from the Arab people by Israeli Zionism should be likewise restored.

The Moro people are fortunate that many Islamic and Arab organizations are extending support and assistance to their revolutionary struggles. The best way to reciprocate these is to fight the enemy firmly and effectively and develop further our self-reliance in the revolutionary struggle.

The Moro people welcome the moral and material support of all peoples. This is an important part of the struggle against imperialism and the local reactionaries. At the same time, the Moro people stand for independence and self-reliance in international affairs.

Should the Moro people be able to establish a people's democratic state ahead of the rest of the Filipino people, they shall develop a foreign policy imbued with the principle of revolutionary internationalism. At the same time, they shall carry out a policy of peaceful co-existence whereby diplomatic trade relations shall be undertaken on the basis of equality and mutual benefit.

Should the Moro people accept regional autonomy in a people's democratic state covering the entire Philippine archipelago, they shall demand a foreign policy no different from what is aforestated.

Mindanao
December 1982

APPENDIX 8

Structure of the Malay Organizations

THE organizational structures described below are constructed on the basis of organization outlines depicted on paper by the respective fronts. Figures A8.1 and A8.2 illustrate the basic structure of the BNPP as specified in its constitution and other documents. The Central Working Committee (Majlis Kerja Tertinggi) of the BNPP is the highest decision-making body when Congress (Majlis Shura) is not in session. It consists of a secretariat which comprises different sections as shown in Figure A8.1. The Central Committee is headed by an elected Chairman and Vice-Chairman and consists of appointed members who hold different portfolios. The BNPP constitution gives power to the Chairman and Vice-Chairman to dissolve the Central Committee and to appoint and dismiss its members. However, they can also be dismissed by a two-thirds majority of the committee members. Both Chairman and Vice-Chairman are elected by a congress which convenes every five years for the purpose. The Congress comprises all members of the Central Committee and two representatives from each local committee and foreign branch (BNPP, 1981a: 12–35; 1981b: 14–15). The BNPP Congress assembles in an Islamic-oriented fashion. For instance, the Congress which was held in August 1981 was saturated with religious rituals. Despite employing a secret ballot, the election result was compromised; one candidate who was elected as Vice-Chairman withdrew from the position in favour of a more suitable individual (the author observed the BNPP Congress held on 4–6 August 1981). Such compromise is said to be characteristic of Islamic-oriented meetings or *shura*. Unlike its counterparts, the BNPP established no permanent tribunal. The tribunal will be created by the Central Committee for a specific case when required (BNPP, 1981a: 14). This suggests that the Front favours self-discipline and believes that those who break the rules are answerable to God.

Figure A8.1
Structure of the BNPP Organization (A)

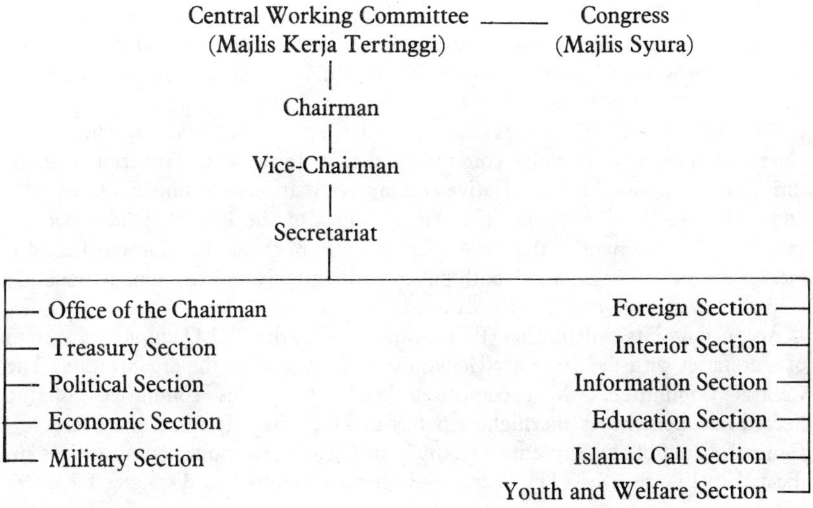

Central Working Committee ——— Congress
(Majlis Kerja Tertinggi)   (Majlis Syura)
            |
         Chairman
            |
       Vice-Chairman
            |
        Secretariat
            |

- Office of the Chairman   Foreign Section
- Treasury Section   Interior Section
- Political Section   Information Section
- Economic Section   Education Section
- Military Section   Islamic Call Section
                                  Youth and Welfare Section

Figure A8.2
Structure of the BNPP Organization (B)

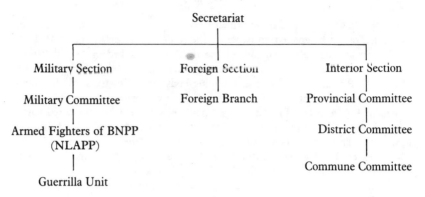

Some sections of the Secretariat consist of several hierarchical units. The Military Section, for instance, extends its authority over the Military Committee which operates the armed *mujahideen* of the BNPP or the National Liberation Army of the Patani People (NLAPP). The NLAPP presently comprises several small units of guerrillas. The Interior Section exercises its jurisdiction over the existing Provincial, District, and Commune Committees in Patani, Yala, and Narathiwat. There is no committee established in Satun Province. The authority of the Foreign Section is over foreign branches including those in Mecca, Cairo, Karachi, Khartoum, and Kuwait. The jurisdiction of these three sections over their lower units, however, applies only in theory. In practice, the Chairman, with a few of his close associates, dominates the overall management of the Front. The concentration of power in the hands of a few leaders is one of the characteristics of the Malay fronts.

Figure A8.3 demonstrates the organizational structure of the BBMP as depicted in its constitution. Unless the Majlis Shura (Congress) is in session, the Presidium is the highest policy-making body of the organization. It can also abrogate decisions of the Congress with the consent of a two-thirds majority of its members. The Presidium consists of the chairmen of the various committees established within the Secretariat (BBMP, n.d.: 4). It appoints and dismisses its own members and all members of the Front. Unlike other fronts, the BBMP has neither elected nor appointed a head of the Presidium; it employs collective leadership. The BBMP consists of a secretariat, Dewan Mujahideen, which comprises different administrative committees that perform various functions. In addition, there are local administrative committees that serve to implement policies and orders of the Presidium. The Majlis Shura of the BBMP, which is composed of the chairmen of the various committees of the secretariat and heads of local committees, functions as both policy-making body and supreme tribunal. It convenes every four months (BBMP, n.d.: 7).

According to its constitution (PULO, n.d.: 3-5), the PULO employs a system of secretariat with the Secretary-General serving as head of the organization. The Central Committee, which comprises heads of various committees of the Secretariat, constitutes the highest policy-making body. It elects the Secretary-General, Chief of the Supreme Tribunal, and heads of various committees as indicated in Figure A8.4. The Secretary-General appoints and dismisses members

Figure A8.3
Structure of the BBMP Organization

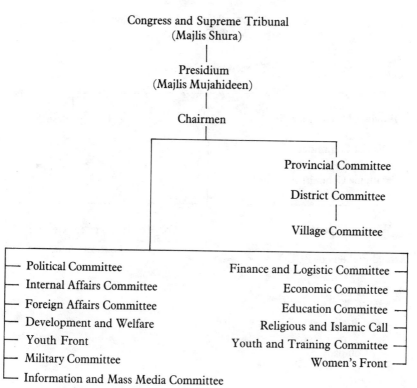

of the Organization, including heads of local committees and foreign branches. The constitution, moreover, authorizes the Secretary-General to dissolve the Central Committee provided that a general election is held within four months. On the other hand, the Secretary-General cannot dismiss those elected members and he can also be dismissed by a two-thirds majority of the Central Committee members.

The essential difference between these organizations is that the BNPP and BBMP are regarded as 'religious-oriented' organizations; both employ Majlis Shura which suggests an Islamic inclination. The PULO, on the other hand, prefers 'secular-orientation'. The BBMP also differs from the rest in that it exercises, at least in theory, collective leadership of a group of chairmen. The BRN does not distribute any documents on the structure of its organization. Its leaders also hesitate to reveal the information when asked. This is partly because the BRN had, in the late 1970s, experienced several arrests of its active leaders, some of whom were given thirty-year sentences in Bangkok. Due to the lack of information, this appendix does not provide the organizational structure of the BRN.

Figure A8.4
Structure of the PULO Organization

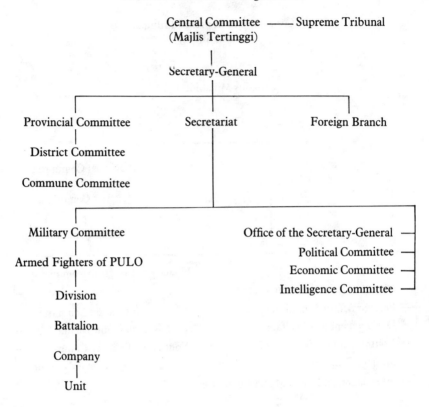

APPENDIX 9

Resolution No. 18 of the political committee at the fifth Islamic Conference of Foreign Ministers held at Kuala Lumpur on June 21–25, 1974

RESOLUTION ON THE PLIGHT OF THE FILIPINO MUSLIMS

PREAMBULAR

The Fifth Islamic Conference of Foreign Ministers held in Kuala Lumpur, Malaysia from 21–25 June 1974:

2. RECALLING Resolution No. 4 adopted by the Fourth Islamic Conference of Foreign Ministers at Benghazi in March 1973;

3. HAVING CONSIDERED the report submitted by the Special Mission composed of the Foreign Ministers of Libya, Saudi Arabia, Senegal and Somalia pursuant to Resolution No. 4 adopted by the Fourth Islamic Conference of Foreign Ministers at Benghazi;

4. EXPRESSING deep appreciation of the efforts of the above-mentioned Special Mission;

5. CONSCIOUS of the complexity of the problem as it relates to an independent and sovereign state but at the same time concerned at the tragic plight of the Filipino Muslims;

OPERATIVE

1. EXPRESSES its deep and continuing anxiety over the situation prevailing among the Filipino Muslims and the Southern Philippines;

2. CALLS upon the Philippine Government to desist from all means which result in the killing of Muslims and the destruction of their properties and places of worship in the Southern Philippines;

3. TAKES NOTE of the steps taken by the Philippine Government to improve the condition of the Muslims, but is convinced that the socio-economic measures proposed by the Philippine Government will not by themselves solve the problems;

4. URGES the Philippine Government to find a political and peaceful solution through negotiations with Muslim leaders, particularly with representatives of the Moro National Liberation Front in order to arrive at a just solution to the plight of the Filipino Muslims within the framework of the national sovereignty and territorial integrity of the Philippines;

5. CALLS ON THE Philippine Government to bring about the necessary climate of confidence for a real and just solution by immediately desisting from military operations, providing protection and security for the Muslims, repatriating refugees to their homes and halting organized Christian migration from the North;

6. APPEALS to peace-loving states and religious international authorities, while recognizing the problem as an internal problem of the Philippines, to use their good offices with the Philippine Government to ensure the safety of Filipino Muslims and the preservation of their liberties in accordance with the Universal Declaration of Human Rights;

7. DECIDES to establish a body to be called the Filipino Muslim Welfare and Relief Agency for the purpose of extending welfare and relief aid direct to Muslims in the Southern Philippines so as to ameliorate their plight and raise social and economic well-being provided the Agency is being financed from the Islamic Fund and is controlled by the Committee of the Solidarity Fund;

8. APPEALS to member countries and others to contribute generously in cash or in kind to the Agency for immediate relief purposes;

9. REQUESTS the Secretary-General in consultation with the four (4) Foreign Ministers to take steps to discuss with the Philippine Government the modalities and terms of cooperation in the work of the Agency;

10. DECIDES that the Special Mission of the Foreign Ministers established in pursuance of Resolution No. 4 of the Fourth Islamic Conference of Foreign Ministers will remain in being to pursue this matter further;

11. DECIDES to review this subject again at the next Islamic Conference of Foreign Ministers;

12. REQUESTS the Secretary-General to transmit this Resolution to the Philippine Government.

APPENDIX 10

Working Paper for the Meeting of the Ministerial Four-member Committee

FIRST—SELF GOVERNMENT

1.(a) In the framework of the national sovereignty of the Philippines and the integrity of its homeland.

Self-government is to be set up in the South in Mindanao, Basilan, Sulu and Palawan. In determining this it was taken into consideration the necessity that the Government of the Philippines put into effect its previous commitments regarding restoring the Southern territories taken away from Muslims after the aggressions which had taken place since war broke out, and land expropriated without any legal right. Territories owned by Muslim ancestors should also be added. Muslims should also go back to the Islamic territories which were owned by Muslims until 1944 the year of the evacuation of Japanese and American Imperialism. All of these territories are to be the geographical area on which Islamic self-government is to be set up.

(b) Christian and non-Christian minorities which have a historical presence on Islamic territories shall remain within the territories under local Islamic Self-Government and shall be considered a Christian minority within Islamic land.

2. When a peace agreement in the Islamic South is concluded, the Muslim Self-Government, which will be the object of this agreement, shall be set up from among good Muslim leaders in the Islamic South, especially the Liberation Front. Other minorities have the right to participate in the Self-Government in the democratic way on the basis of the principle of similar treatment for Muslims living in other territories in the Republic of the Philippines.

SECOND—AFFAIRS OF INTERNAL SECURITY IN THE ISLAMIC TERRITORIES

Units of local security, defense and militia shall be set up basically from among good elements of the Liberation Front and Muslim leaderships in the South, also with the participation of other minorities.

THIRD—DEFENSE AND FOREIGN POLICY

1. National Security and central defense are the concern of the Central Government and the Muslims through participation in the central Armed Forces and in Foreign policy as a national duty.

2. Dissolving all units of the special force of the Philippine Army and their bases as well as armed units set up on account of the war in the South.

FOURTH—ADMINISTRATIVE SYSTEM

The government of the Muslim territory undertake all internal administrative matters. The Central Government shall supervise all national administrative matters in the territory and co-ordinate the efforts of the two authorities in matters of common interest as well as matters which are an object of conflict.

FIFTH—SYSTEM OF COURTS

1. Local and central courts shall be set up. In case of conflict in specialization the matter shall be referred to the Supreme Legal Authority.

2. Local law courts shall have jurisdiction in local cases related to the territory in accordance with the principles of Islamic *Sharia*.

3. A Supreme Court shall be set up from Eight neutral judges characterized by their integrity, four of whom shall be selected by the Central Authority and the other four by the local authority. Chairmanship of the Court shall be alternate by a member of any of the two sides. Their appointments shall be by a presidential decree. This Court shall be concerned with:

(a) Law suits related to administrative matters in case of conflict of terms of reference; (b) conflict of terms of reference between the central law court and local law courts; (c) any conflict referred to it by local or central government.

SIXTH—SYSTEM OF EDUCATION

1. Educational institutions up to the secondary schools shall be controlled by the self-government of the territory. Co-ordination of higher and university education shall take place with the Central Government.

2. Rebuilding the Islamic educational system in the territory in co-ordination with the general national educational system.

SEVENTH—ESTABLISHMENT OF ISLAMIC LIFE AND SOCIETY IN THE SOUTH

The practice of Muslim religious rites including pilgrimage, and the supervision and administration of mosques, places of worship and Islamic institutions shall fall within the jurisdiction of the territory's government.

EIGHTH—FINANCIAL AND ECONOMICAL AFFAIRS

1. The territory shall have the right to run internal economic affairs and to set up the financial sources of the territory. The territory's government shall have the right to impose and collect taxes. Its tax commitments shall be paid directly to the Central Government in return for the latter's undertaking of its duty towards development projects in the territory for the realization of the rebuilding of the South and cultural and economic development.

2. The Central Government shall extend the necessary financial support to the territory's government to enable it to shoulder its responsibilities in public services within the general budget with special emphasis on the territory due to the destruction to which it was exposed.

3. Finances devoted to development projects which are received by the territory from Islamic states shall be invested and managed through co-ordination between the two authorities. A joint council shall be set up which shall submit periodical progress reports to the Secretary General of the Islamic Conference on projects financed by Islamic states.

NINTH—Inhabitants of the territory, in their capacity as Philippine citizens, shall have the right to participate in the Central Government and all organs of the state.

APPENDIX 11

Resolution No. 10 of the Political Committee at the sixth Islamic Conference of Foreign Ministers

RESOLUTION ON THE NEGOTIATIONS BETWEEN THE MORO LIBERATION FRONT AND THE GOVERNMENT OF THE PHILIPPINES

The Sixth Islamic Conference of Foreign Ministers meeting in Jeddah from 3 to 7 Rajab 1395 H. corresponding to 12 to 15 July 1975.

HAVING examined the plan of action prepared by the Committee of Four which [is] composed of the Foreign Ministers of the Republic of Senegal, the Kingdom of Saudi Arabia, the Libyan Arab Republic and Democratic Republic of Somalia, in accordance with Resolution 4 of the Fourth Islamic Conference and in pursuance of Resolution 5 of the Fifth Islamic Conference;

EXPRESSES its satisfaction at the effort of the Ministerial Committee of Four and decides that the Committee should pursue its appointed tasks in accordance with its terms of reference until such time as complete agreement is reached between the Government of the Philippines and the Moro Liberation Front;

APPROVES the plan of action prepared by the Ministerial Committee of Four and considers that this plan is the fundamental basis for any settlement of the problem, in such manner as would be in keeping with Muslim objectives for autonomy within the territorial integrity of the Philippines, in accordance with Resolution 18 of the Fifth Islamic Conference of Kuala Lumpur;

NOTES with satisfaction that the Moro Liberation Front has agreed to the plan of action submitted as the fundamental basis for negotiations and expresses satisfaction at the initiative of the Government of the Philippines to accept autonomy for Muslims in Mindanao, Basilan, Sulu and Palawan;

ENTRUSTS the Ministerial Committee of Four and the Secretary-General with the task of contacting the Government of the Philippines to invite them to the negotiations as a basis of the afore-mentioned plan of action, in a manner designed to achieve agreement on all aspects of self-government for the Muslims in the South of the Philippines, ensuring peace and security to them and guaranteeing all their legitimate rights, while at the same time preserving the territorial integrity of the Philippines, thus enabling the government of the latter to exert all their efforts in the service of the people as a whole;

REQUESTS the Secretary-General in consultation with the Foreign Ministers, to take all necessary steps for holding such negotiations within the shortest possible time at the Headquarters of the General Secretariat in Jeddah;

DECIDES to review the results achieved so far by the two parties during the next Islamic Conference of Foreign Ministers;

REQUESTS the Secretary-General to communicate this resolution to the Government of the Philippines and to the Moro Liberation Front.

Bibliography

Abbahil, Abdulsiddik A. (1983), 'The Bangsa Moro: Their Self-image and Intergroup Ethnic Attitudes', Master's thesis, San Carlos University-Dansalan Research Centre Consortium.
Abbas, M. Yahya (1979), 'Developments in Moro Struggle', in Alfredo T. Tiamson and Rosalinda N. Caneda (comps.), *The Southern Philippines Issue: Readings in the Mindanao Problem*, Twelfth Annual Seminar on Mindanao–Sulu Culture, Mindanao State University.
Abdullah Al-Qari Haji Salleh (1974), 'To' Kenali: His Life and Influence', in William R. Roff (ed.), *Kelantan*, Kuala Lumpur: Oxford University Press.
Abubakar, Carmen A. (1984), 'The Religious Dimension of the Moro Problem: A Restatement', in Christian Conference of Asia, *Religion and Asian Politics: An Islamic Perspective Report of the Consultation on Religion and Asian Politics*, Hong Kong: Christian Conference of Asia.
AFRIM Resource Center (1980), *Mindanao Report: A Preliminary Study on the Economic Origins of Social Unrest*, Davao City: AFRIM Resource Center.
Ahmad, Aijas (1980), 'Class and Colony in Mindanao: Political Economy of the "National Question"', mimeographed.
Al-Attas, S. Naguib (1969), *A General Theory of the Islamization of the Malay–Indonesian Archipelago*, Kuala Lumpur: Dewan Bahasa dan Pustaka.
Algar, Hamid (1983), *The Roots of the Islamic Revolution*, London: The Open Press.
Ali, A. Yusuf (trans.) (1983), *The Holy Qur'an*, Maryland: Amana Corp.
Alonto, Abducal W. (1983), 'The Influence of Traditional Agama Leaders on the Decision-making of the Barangay Council', Master's thesis, Mindanao State University.
Alonto, Abdul Ghafur Madki (1982), 'Management and Organization of Madrasah', *FEPE Review*, 12(3 and 4): 31–3.
Alonto, Ahmad Domocao (1979), 'Conspiracy to Liquidate Islam in the Philippines—II', *The Journal, Rabitat Al-Alam Al-Islami*, 6(11): 59–62.
Andaya, Barbara W. and Andaya, Leonard Y. (comps.) (1982), *A History of Malaysia*, London: The Macmillan Press.
Anderson, Benedict (1983), *Imagined Communities: Reflections on the Origin and Spread of Nationalism*, London: Verso Editions and New Left Books.
Anderson, Charles W., von der Mehden, Fred R., and Young, Crawford (1967), *Issues of Political Development*, New Jersey: Prentice-Hall.
Ansar El-Islam (1974), *Manifesto*, presented by Ansar El-Islam at the opening session of the Muslim Conference on Government Policies and Programs, Mindanao State University, Marawi City.
Apter, David A. (1967), *The Politics of Modernization*, Chicago: University of Chicago Press.
Arce, Wilfredo F. (1983), *Before the Secessionist Storm: Muslim–Christian Politics*

in Jolo, Sulu, Philippines 1961–62, Singapore: Maruzen Asia.

Aruri, Naseer H. (1977), 'Nationalism and Religion in the Arab World: Allies or Enemies', *Muslim World*, 67(4): 266–79.

Asani, Abdurasad (1985), 'The Bangsamoro People: A Nation in Travail', *Journal, Institute of Muslim Minority Affairs*, 6(2): 295–314.

Ayoob, Mohammed (1984), 'Concluding Discussion', in Lim Joo-Jock and Vani S. (eds.), *Armed Separatism in Southeast Asia*, Singapore: Institute of Southeast Asian Studies.

_____ (ed.) (1981), *The Politics of Islamic Reassertion*, London: Croom Helm.

Bangnara, A. (1976), *Patani: Past-Present*, Bangkok: Chomrom Saengtian (in Thai).

Bangsa Moro Liberation Organization (BMLO) (1978), A Policy Statement issued by the Supreme Executive Council of the BMLO, 19 May.

Bangsa Muslimin Islamic Liberation Organization (BMILO) (1984a), *The First Declaration*, declaration issued in October by the BMILO Executive Council, Mecca.

_____ (1984b), *The Second Declaration*, declaration issued in December by the BMILO Executive Council, Mecca.

Banton, Michael (1983), *Racial and Ethnic Competition*, Cambridge: Cambridge University Press.

Barisan Bersatu Mujahideen Patani (BBMP) (n.d.), The Constitution of the BBMP distributed to its members, n.p. (in Malay).

_____ (1986), *Conference of Friendly Leaders: Parties, Organizations, and Movements in Patani*, letter of invitation to a conference of leaders of various fronts held in Kuala Lumpur on 13–14 June (in Malay).

Barisan Nasional Pembebasan Patani (BNPP) (1976), *The Muslim Struggle for Survival in South Thailand*, document distributed at the Seventh Conference of Foreign Ministers held in May 1976 in Istanbul.

_____ (1977), *Question of the Human Rights of Persons Subjected to Any Form of Detention or Imprisonment*, letter of complaint, dated 18 February 1977, submitted to the United Nations Office at Geneva through the Muslim World League, Mecca.

_____ (1980), *Memorandum on Patani Muslim Struggle*, document submitted to the Eleventh Islamic Conference of Foreign Ministers held in April 1980 in Islamabad.

_____ (1981a), *The National Liberation Front of Patani*, Patani: Information Bureau.

_____ (1981b), *BNPP Constitution*, Patani: Information Bureau (in Malay).

_____ (1981c), *Programmes of the BNPP's Second General Meeting and Reports of Its Past Activities*, paper presented by the BNPP at the BNPP's Second General Meeting, 4–6 August 1981 (in Malay).

_____ (1981d), *Programmes of Education*, pamphlet distributed by the BNPP to its members (in Malay).

_____ (1982), *Monthly Contributions to BNPP*, letter distributed by the BNPP's Economic Section to members of the Central Working Committee (in Malay).

Barth, Frederik (1969), *Ethnic Groups and Boundaries: The Social Organization of Culture Difference*, Boston: Little, Brown.

Bastin, John and Winks, Robin W. (1979), *Malaysia: Selected Historical Readings*, Nendeln: KTO Press.

Bates, Robert H. (1974), 'Ethnic Competition and Modernization in Contemporary Africa', *Comparative Political Studies*, 6: 457–84.

Bayat, Margol (1980), 'Islam in Pahlavi and Post-Pahlavi Iran: A Cultural-

Revolution?', in John L. Esposito (ed.), *Islam and Development: Religion and Sociopolitical Change*, New York: Syracuse University Press.

Beckett, Jeremy (1982), 'The Defiant and the Compliant: The Datus of Magindanao under Colonial Rule', in Alfred McCoy and C. de Jesus (eds.), *Philippine Social History: Global Trade and Local Transformations*, Quezon City: Ateneo de Manila University Press.

Bennigsen, Alexandre and Broxup, Marie (1983), *The Islamic Threat to the Soviet State*, London and Canberra: Croom Helm.

Bennigsen, Alexandre and Lemercier-Quelquejay, Chantal (1967), *Islam in the Soviet Union*, London: Pall Mall Press.

Bennigsen, Alexandre A. and Wimbush, S. Enders (1979), *Muslim National Communism in the Soviet Union: A Revolutionary Strategy for the Colonial World*, Chicago: University of Chicago Press.

Bentley, George C. (1978), 'Historical Perspective on the Muslim Armed Struggle? (Critical Review of Samuel Tan's *The Filipino Muslim Armed Struggle, 1900–1972*)', *Mindanao Journal*, 5(2): 135–53.

_____ (1982), 'Law, Disputing, and Ethnicity in Lanao, Philippines', Doctoral dissertation, University of Washington.

Birch, Anthony H. (1978), 'Minority Nationalist Movements and Theories of Political Integration', *World Politics*, 30(3): 325–44.

Bonacich, Edna (1972), 'A Theory of Ethnic Antagonism: The Split Labor Market', *American Sociological Review*, 37: 547–59.

_____ (1976), 'Advanced Capitalism and Black/White Relations', *American Sociological Review*, 41: 31–51.

_____ (1979), 'The Past, Present and Future of Split Labor Market Theory', *Research in Race and Ethnic Relations*, 1: 17–64.

Bonacich, Edna and Modell, John (1980), *The Economic Basis of Ethnic Solidarity*, Berkeley: University of California Press.

Bonney, R. (1971), *Kedah 1771–1821: The Search for Security and Independence*, Kuala Lumpur: Oxford University Press.

Bottomore, T. B. (1964), *Elites and Society*, Harmondsworth: Penguin Books.

Boulding, Elise (1979), 'Ethnic Separatism and World Development', in Louis Kriesberg (ed.), *Research in Social Movements, Conflict and Change*, Connecticut: JAI Press.

Bresnan, John (ed.) (1986), *Crisis in the Philippines: The Marcos Era and Beyond*, New Jersey: Princeton University Press.

Bruno, Juanito A. (1973), *The Social World of the Tausug*, Manila: Centro Escolar University Research and Development Centre.

Bunnag, Tej (1971), *Revolt in the Seven Provinces in 1902*, Bangkok: Thai Watthanapanich (in Thai).

_____ (1977), *The Provincial Administration of Siam, 1892–1915: The Ministry of of the Interior under Prince Damrong Rajanubhab*, Kuala Lumpur: Oxford University Press.

Burg, Steven L. (1984), 'Muslim Cadres and Soviet Political Development: Reflection from a Comparative Perspective', *World Politics*, 37(1): 24–47.

Calvert, Peter (1970), *Revolution*, London: Macmillan.

Casanova, Pablo G. (1963), 'Internal Colonialism and National Development', *Studies in Comparative International Development*, 1(3): 27–37.

Centre for Administration of the Southern Border Provinces (n.d.), *Notice: Activities of PULO in Saudi Arabia*, pamphlet distributed in the four Muslim provinces (in Thai).

Che Man, W. K. (1983), 'Muslim Elites and Politics in Southern Thailand', Master's thesis, Universiti Sains Malaysia.

_____ (1985), 'The Malay-Muslims of Southern Thailand', *Journal, Institute of Muslim Minority Affairs*, 6(1): 98–112.
Cohen, Abner (1969), *Custom and Politics in Urban Africa*, London: Routledge & Kegan Paul.
Coleman, James S. (1960), 'The Politics of Sub-Saharan Africa', in Gabriel A. Almond and James S. Coleman (eds.), *The Politics of the Developing Areas*, New Jersey: Princeton University Press.
Communist Party of the Philippines (CPP) (1969), 'Constitution of the Communist Party of the Philippines', in Alfredo B. Saulo, *Communism in the Philippines: An Introduction*, Manila: Ateneo de Manila University.
Connor, Walker (1967), 'Self-Determination: the New Phase', *World Politics*, 20(1): 30–53.
_____ (1969), 'Ethnology and the Peace of South Asia', *World Politics*, 22(1): 51–86.
_____ (1972), 'Nation-Building or Nation-Destroying?', *World Politics*, 24(3): 319–55.
Cudsi, Alexander and Dessouki, Ali H. (eds.) (1981), *Islam and Power*, London: Croom Helm.
Dahrendorf, Ralf (1969), *Class and Class Conflict in Industrial Society*, Stanford: Stanford University Press.
Darwin, Charles (1979), *The Illustrated Origin of Species*, New York: Hill & Wang.
David, Virgilio (1977), 'Barriers in the Development of the Coconut Industry', Master's thesis, Ateneo de Manila Graduate School of Business.
De Vos, George and Romanucci-Ross, Lola (eds.) (1975), *Ethnic Identity: Cultural Continuities and Change*, California: Mayfield Publishing Company.
Despres, Leo (ed.) (1975), *Ethnicity and Resource Competition in Plural Societies*, The Hague: Mouton Publishers.
Deutsch, Karl W. (1961), 'Social Mobilization and Political Development', *American Political Science Review*, 55(3): 493–514.
_____ (1963), 'Nation-Building and National Development: Some Issues of Political Research', in Karl Deutsch and William Foltz (eds.), *Nation-Building*, New York: Atherto Press.
_____ (1966), *Nationalism and Social Communication*, Massachusetts: The MIT Press.
_____ (1969), *Nationalism and Its Alternatives*, New York: Knoft.
_____ (1970), *Politics and Government*, Boston: Houghton Mifflin.
Dos Santos, Theotonio (1970), 'The Structure of Dependence', *American Economic Review*, 60: 231–6.
Dulyakasem, Uthai (1981), 'Education and Ethnic Nationalism: A Study of the Muslim-Malays in Southern Siam', Doctoral dissertation, Stanford University.
_____ (1984), 'Muslim-Malay Separatism in Southern Thailand: Factors Underlying the Political Revolt', in Lim Joo-Jock and Vani S. (eds.), *Armed Separatism in Southeast Asia*, Singapore: Institute of Southeast Asian Studies.
Dumarpa, Jaime T. (1983), 'An Exploratory Study of Maranao Muslims' Concepts of Land Ownership: Its Implications for the Mindanao Conflict', Master's thesis, University of San Carlos, Cebu City.
Duverger, Maurice (1976), *Political Parties*, Cambridge: Methuen.
Emerson, Rupert (1960), *From Empire to Nation: The Rise of Self Assertion of Asian and African Peoples*, Massachusetts: Harvard University Press.
Enloe, Cynthia (1973), *Ethnic Conflict and Political Development*, Boston: Little, Brown & Company.
Esman, Milton J. (ed.) (1977), *Ethnic Conflict in the Western World*, Ithaca: Cornell University Press.

Esposito, John L. (ed.) (1980), *Islam and Development: Religion and Sociopolitical Change*, New York: Syracuse University Press.

Europa Publications (1985), *The Middle East and North Africa 1986*, London: Europa Publications.

Fanon, Frantz (1965), *A Dying Colonialism*, New York: Grove Press.

Faruki, Kemal A. (1983), 'Ethnic Resurgence in the West', typescript.

Federspiel, Howard M. (1985), 'Islam and Development in the Nations of ASEAN', *Asian Survey*, 25(8): 805–21.

Filipinas Foundation (1971), *An Anatomy of Philippine Muslim Affairs*, Manila: Filipinas Foundation.

Forbes, W. Cameron (1945), *The Philippine Islands*, Massachusetts: Harvard University Press.

Fraser, Thomas M. (1960), *Rusembilan: A Malay Fishing Village in Southern Thailand*, New York: Cornell University Press.

—— (1966), *Fishermen of South Thailand: The Malay Villagers*, New York: Holt & Winston.

Gabongan Melayu Patani Raya (GAMPAR) (1948), *Some Facts about Malays in South Siam*, Kota Bharu: Information Bureau.

Geertz, Clifford (1960), *The Religion of Java*, Illinois: The Free Press of Glencoe.

—— (1963), 'The Integrative Revolution: Primordial Sentiment and Civil Politics in the New States', in Clifford Geertz (ed.), *Old Societies and New States*, New York: The Free Press.

—— (1968), *Islam Observed: Religious Development in Morocco and Indonesia*, Chicago: University of Chicago Press.

Gellner, Ernest (1973), 'Scale and Nation', *Philosophy of the Social Sciences*, 3: 1–17.

George, T. J. S. (1980), *Revolt in Mindanao: The Rise of Islam in Philippine Politics*, Kuala Lumpur: Oxford University Press.

Gerth, Hans C. and Mills, C. Wright (1953), *Character and Social Structure*, New York: Harcourt, Brace & Company.

Gibb, H. A. R. and Kramers, J. H. (eds.) (1953), *Shorter Encyclopaedia of Islam*, Leiden: E. J. Brill.

Glang, Alunan C. (1969), *Muslim Secession or Integration?*, Quezon City: R. P. Garcia.

—— (1973), 'Realities and Illusions in the Muslim Conflict', *The National Security Review*, 1(4): 34–69.

—— (1974), 'Modernizing the Muslims', in Peter G. Gowing and Robert D. McAmis (eds.), *The Muslim Filipinos*, Manila: Solidaridad Publishing House.

—— (1976), *Briefing and Debriefing on Moro Liberation Front Leadership at the 7th Islamic Conference of Foreign Ministers for Civil Intelligence and Security Authority (CISA) of the National Intelligence and Security Authority (NISA)*, report submitted to the NISA, 15 July.

Glazer, Nathan and Moynihan, Daniel P. (eds.) (1975), *Ethnicity: Theory and Experience*, Massachusetts: Harvard University Press.

Gowing, Peter G. (1962), 'Resurgent Islam and the Moro Problem in the Philippines', *South East Asia Journal of Theology*, 4(1): 57–65.

—— (1969), 'How Muslim are the Muslim Filipinos?', *Solidarity*, 4(8): 21–9.

—— (1975), 'Moro and Khaek: The Position of Muslim Minorities in the Philippines and Thailand', *Southeast Asian Affairs*, Singapore: Institute of Southeast Asian Studies.

—— (1977), *Mandate in Moroland: The American Government of Muslim*

Filipinos 1899–1920, Quezon City: New Day Publishers.

——— (1979), *Muslim Filipinos—Heritage and Horizon*, Quezon City: New Day Publishers.

——— (1981), 'America's Proconsuls in Mindanao, 1899–1913', *Dansalan Quarterly*, 3(1): 5–28.

Greene, Thomas H. (1974), *Comparative Revolutionary Movements*, New Jersey: Prentice-Hall.

Gullick, J. M. (1958), *Indigenous Political Systems of Western Malaya*, London: The Athlone Press.

——— (1981), *Malaysia: Economic Expansion and National Unity*, London: Westview Press.

Haddad, Yvonne (1980), 'The Arab-Israeli Wars, Nasserism, and the Affirmation of Islamic Identity', in John Esposito (ed.), *Islam and Development: Religion and Sociopolitical Change*, New York: Syracuse University Press.

Haemindra, Nantawan (1976), 'The Problem of the Thai-Muslims in Four Southern Provinces of Thailand', Part 1, *Journal of Southeast Asian Studies*, 7(2): 197–225.

——— (1977), 'The Problem of the Thai-Muslims in Four Southern Provinces of Thailand', Part 2, *Journal of Southeast Asian Studies*, 8(1): 85–105.

Hakim, Khalifa A. (1974), *Islamic Ideology*, Lahore: Institute of Islamic Culture.

Hall, D. G. E. (1981), *A History of South-East Asia*, New York: St. Martin's Press.

Hall, Raymond L. (ed.) (1979), *Ethnic Autonomy—Comparative Dynamics: The Americas, Europe and the Developing World*, New York: Pergamon Press.

Hannan, Michael T. (1979), 'The Dynamics of Ethnic Boundaries in Modern States', in Michael T. Hannan and John Meyer (eds.), *National Development and World System: Educational, Economic and Political Change, 1950–1970*, Chicago: University of Chicago Press.

Harrison, Francis Burton (1922), *The Corner-Stone of Philippine Independence: A Narrative of Seven Years*, New York: The Century Co.

Hashim, Salamat (1977), Letter (dated 24 December) from Salamat Hashim, Chairman of the MNLF-Hashim Faction, to Dr Ahmadu Karim Gaye, Secretary-General of the Organization of the Islamic Conference, Jeddah.

——— (1985), *The Bangsamoro Mujahid: His Objectives and Responsibilities*, Mindanao: Bangsamoro Publications.

Hassan, Abduljim J. (1978), 'Characteristics of Backdoor Migrants to Sabah, Malaysia from the Philippines: The Case of the Two Island Communities of Tawi-Tawi', *Mindanao Journal*, 5(2): 98–119.

Hassan, Hatimil (1981), 'The Moro National Liberation Front and the Present Resistance', in Komite ng Sambayanang, *Philippines: Repression and Resistance*, Komite ng Sambayanang, Pilipino.

Hechter, Michael (1973), 'The Persistence of Regionalism in the British Isles, 1885–1966', *American Journal of Sociology*, 79(2): 319–42.

——— (1974), 'The Political Economy of Ethnic Change', *American Journal of Sociology*, 79(5): 1151–78.

——— (1975), *Internal Colonialism*, London: Routledge & Kegan Paul.

——— (1978), 'Group Formation and Cultural Division of Labor', *American Journal of Sociology*, 84(2): 293–318.

Hermosa, Jesus R. (1980), 'The Mindanao Conflict', *The National Security Review*, 7(1): 3–19.

Hilton, Mike (1979), 'The Split Labor Market and Chinese Immigration, 1848–1882', *Journal of Ethnic Studies*, 6(4): 99–108.

Hitchner, Dell and Levine, Carol (1973), *Comparative Government and Politics*, New York: Dodd, Mead & Company.

Hooton, Earnest A. (1947), *Up From the Ape*, New York: The Macmillan Company.

Horowitz, Donald L. (1971), 'Multiracial Politics in the New States: Toward a Theory of Conflict', in Robert J. Jackson and Michael B. Stein (eds.), *Issues in Comparative Politics*, New York: St Martin's Press.

_____ (1985), *Ethnic Groups in Conflict*, Berkeley: University of California Press.

Huntington, Samuel P. (1971), 'The Change to Change: Modernization, Development, and Politics', *Comparative Politics*, 3(3): 283–322.

Hussin, Parouk (1981), 'The Marcos Regime Campaign of Genocide', in Komite ng Sambayanang, *Philippines: Repression and Resistance*, Komite ng Sambayanang, Pilipino.

Ibrahim Shukri (n.d.), *History of the Malay Kingdom of Patani*, Kelantan: Majlis Ugama Islam Press (in Malay).

Isidro, Antonio (1979), *The Moro Problem: An Approach through Education*, Marawi City: University Research Center, Mindanao State University.

Islamic Conference of Foreign Ministers (ICFM) (1974), *Resolution No. 18: The Plight of the Filipino Muslims*, Resolution adopted by the Second Islamic Summit held on 22–24 February in Lahore, Pakistan.

_____ (1977a), *Resolution No. 7/8-P: The Question of Muslims in South Philippines*, Resolution adopted by the Eighth Islamic Conference of Foreign Ministers held on 16–22 May in Tripoli, Libya.

_____ (1977b), *Resolution No. 2/8-P: Granting, as an Exceptional Measure, the Status of Observer to the Moro National Liberation Front*, Resolution adopted by the Eighth Islamic Conference of Foreign Ministers held on 16–22 May in Tripoli, Libya.

Israeli, Raphael (1978), *Muslims in China: A Study in Cultural Confrontation*, London: Curzon Press.

Jahan, Rounag (1972), *Pakistan: Failure in National Integration*, New York: Columbia University Press.

Jansen, G. H. (1979), *Militant Islam*, London: Pan Books.

Jubair, Salah (1984), *Bangsamoro: A Nation Under Endless Tyranny*, Lahore: Islamic Research Academy.

Kautsky, John H. (1971), 'Nationalism', in H. G. Kebschull (ed.), *Politics in Transitional Societies*, New York: Appleton-Century-Crofts.

Kerkvliet, Benedict J. (1977), *The Huk Rebellion: A Study of Peasant Revolt in the Philippines*, Berkeley: University of California Press.

Kershaw, Roger (1985), 'Nationalists or Buddhists? The Response of Thai Legislators to a Case of Perceived Malaysian Interference in the South in 2517 (1974)', paper presented at Asian Regional Workshop on Ethnic Minorities in Buddhist Polities held on 25–28 June 1985 at Chulalongkorn University.

Kettani, M. Ali (1986), *Muslim Minorities in the World Today*, London: Mansell Publishing.

Khadduri, Majid (1964), 'The Islamic Philosophy of War', in Joel Larus (ed.), *Comparative World Politics*, Belmont, California: Wadsworth Publishing Company.

Khan, Inamullah (1979), 'The Situation in Southern Philippines', typescript.

Khan, Qamaruddin (1973), *The Political Thought of Ibn Taymiyah*, Islamabad: Islamic Research Institute.

Khomeini, Imam (1981), *Islam and Revolution*, Berkeley: Mizan Press.

Kiefer, Thomas M. (1972a), 'The Tausug Polity and the Sultanate of Sulu: A Segmentary State in the Southern Philippines', in Gerard Rixhon (ed.), *Sulu Studies 1*, Jolo: Notre Dame of Jolo College.

―――― (1972b), *The Tausug: Violence and Law in a Philippine Muslim Society*, New York: Holt, Rinehart & Winston.

Koch, Margaret L. (1977), 'Patani and the Development of a Thai State', *Journal of the Malaysian Branch of the Royal Asiatic Society*, 50(2): 69–88.

Kohn, Hans (1967a), *The Idea of Nationalism*, New York: Collier-Macmillan.

―――― (1967b), *Prelude to Nation-States: The French and German Experience, 1789–1815*, New York: Van Nostrand.

Komite ng Sambayanang (1981), *Philippines: Repression and Resistance*, Komite ng Sambayanang, Pilipino.

Krejci, Jaroslav and Velimsky, Vitezslav (1981), *Ethnic and Political Nations in Europe*, London: Croom Helm.

Laffin, John (1979), *The Dagger of Islam*, London: Sphere Books.

Lapidus, Gail W. (1984), 'Ethnonationalism and Political Stability: The Soviet Case', *World Politics*, 36(4): 555–80.

Lasswell, Harold D. and Lerner, Daniel (eds.) (1965), *World Revolutionary Elites: Studies in Coercive Ideological Movements*, Massachusetts: The MIT Press.

Leenanon, Harn (1984), *Political Ideology*, Bangkok: Samnakpim Suesan (in Thai).

Leifer, Eric M. (1981), 'Competing Models of Political Modernization: The Role of Ethnic Ties', *American Journal of Sociology*, 87(1): 23–47.

Lenski, Gerhard (1961), *The Religious Factor: A Sociological Study of Religion's Impact on Politics, Economics and Family Life*, New York: Doubleday.

Lewy, Guenter (1974), *Religion and Revolution*, New York: Oxford University Press.

Lieberson, Stanley (1970), *Language and Ethnic Relations in Canada*, New York: John Wiley & Sons.

Lijphart, Arend (1977), 'Political Theories and Explanation of Ethnic Conflict in the Western World: Falsified Predictors and Plausible Predictions', in Milton J. Esman (ed.), *Ethnic Conflict in the Western World*, Ithaca: Cornell University Press.

Linz, Juan (1973), 'Early State-building and Late Peripheral Nationalisms against the State: The Case of Spain', in S. N. Eisenstadt and Stein Rokkan (eds.), *Building States and Nations*, Beverly Hills: Sage Publications.

Lipset, Seymour M. and Rokkan, Stein (1967), *Party Systems and Voter Alignments*, New York: The Free Press.

Lucman, Sultan Harun Al Rashid (1982), *The Chairman's Message to the Bangsa Moro People for Unity*, n.p.: Bangsa Moro Liberation Organization.

Lynch, Frank (1959), *Social Class in a Bikol Town*, Chicago: University of Chicago Philippine Studies Program.

M. Noordin Sopiee (comp.) (n.d.), *The South Siam Secession Movement and the Battle for Unification with Malaya: A Historical Sourcebook*, n.p.

Madale, Nagasura T. (1976), 'A Look at Philippine Mosques', *Salaam*, 3(1): 12–14.

―――― (1984), 'The Future of the MNLF as a Separatist Movement in Southern Philippines', in Lim Joo-Jock and Vani S. (eds.), *Armed Separatism in Southeast Asia*, Singapore: Institute of Southeast Asian Studies.

Majul, Cesar A. (n.d.), *Muslims in the Philippines: Past, Present, and Future Prospects*, n.p.

_____ (1962), 'Theories on the Introduction and Expansion of Islam in Malaysia', *Association of Historians of Asia*, 2nd Biennial Conference.

_____ (1964), 'Political and Historical Notes on the Old Sulu Sultanate', Paper No. 30, International Conference on Asian History, University of Hong Kong.

_____ (1970), 'Islamic Influences in the Philippines', paper read before a meeting of the Ansar El-Islam, Marawi City.

_____ (1973), *Muslims in the Philippines*, Quezon City: The University of the Philippines Press.

_____ (1974), 'The Muslims in the Philippines: An Historical Perspective', in Peter G. Gowing and Robert D. McAmis (eds.), *The Muslim Filipinos*, Manila: Solidaridad Publishing House.

_____ (1977), 'Mosques in the Philippines', in Lahing Pilipino Publishing, *Filipino Heritage*, Manila: Lahing Pilipino Publishing.

Makari, Victor E. (1983), *Ibn Taymiyyah's Ethics: The Social Factor*, Chico, California: Scholars Press.

Mao Tse-Tung and Guevara, Che (1962), *Guerrilla Warfare*, London: Cassell.

Marx, Karl (1964), *Pre-capitalist Economic Formations* (trans. by Jack Cohen), New York: International Publishers.

_____ (1969), *On Colonialism and Modernization* (edited by Shlomo Avineri), New York: Doubleday.

Mastura, Michael O. (1980), *Islam and Development: A Collection of Essays by Cesar Adib Majul*, Manila: OCIA Publications.

_____ (1982), 'Assessing the Madrasah as an Educational Institution: Implications for the Ummah', *FAPE Review*, 12(3 and 4): 6–15.

_____ (1984), 'Development Programs for Mindanao and Sulu: Retrospect and Prospect', in Datu Michael O. Mastura, *Muslim Filipino Experience: A Collection of Essays*, Manila: Ministry of Muslim Affairs.

May, R. J. (1981), 'The Philippines', in Mohammed Ayoob (ed.), *The Politics of Islamic Reassertion*, London: Croom Helm.

_____ (1984), 'The Situation of Philippine Muslims', *Journal, Institute of Muslim Minority Affairs*, 5(2): 427–39.

_____ (1985), 'Muslim and Tribal Filipinos', in R. J. May and Francisco Nemenzo (eds.), *The Philippines After Marcos*, London: Croom Helm.

May, R. J. and Nemenzo, Francisco (eds.) (1985), *The Philippines After Marcos*, London: Croom Helm.

McAmis, Robert D. (1974), 'Muslim Filipinos: 1970–1972', in Peter G. Gowing and Robert D. McAmis (eds.), *The Muslim Filipinos*, Manila: Solidaridad Publishing House.

McCarthy, John and Zald, Mayer (1977), 'Resource Mobilization in Social Movements: A Partial Theory', *American Journal of Sociology*, 82: 1212–39.

McVey, Ruth (1984), 'Separatism and the Paradoxes of the Nation-State in Perspective', in Lim Joo-Jock and Vani S. (eds.), *Armed Separatism in Southeast Asia*, Singapore: Institute of Southeast Asian Studies.

Mednick, Melvin (1957), 'Some Problems of Moro History and Political Organization', *Philippine Sociological Review*, 5(1): 39–52.

Megarat, Manas (1977), 'The Failure in Subjugating Terrorists of the Three Southern Provinces', *Research Document*, syllabus for police officer course, Group 13 (in Thai).

Melchor, Alejandro (1973), 'Reconstruction and Development Program for Mindanao—A Summary', in PTF-RDM, *Report on the Reconstruction and*

Development Program for Mindanao, Manila: Presidential Task Force for the Reconstruction and Development of Mindanao.

Melson, Robert and Wolpe, Howard (1970), 'Modernization and the Politics of communalism: A Theoretical Perspective', *American Political Science Review*, 64: 1112-30.

Mercado, Eliseo (1981), 'The Moro Contemporary Armed Struggle—A Jihad?', Master's thesis, Pontifical Gregorian University, Rome.

Meyer, John W. and Hannan, Michael T. (eds.) (1979), *National Development and the World System*, Chicago: University of Chicago Press.

Michels, Robert (trans.) (1915), *Political Parties: A Sociological Study of the Oligarchical Tendencies of Modern Democracy*, London: Jarrolds & Sons.

Miller, Robert F. (1986), A book review, 'Romuald J. Misiunas and Rein Taagepara, *The Baltic States: Years of Dependence, 1940–1980*', *Australian Outlook*, 40(1): 56-7.

Miller, Stuart C. (1982), *'Benevolent Assimilation': The American Conquest of the Philippines, 1899–1903*, New Haven: Yale University Press.

Milne, R. S. (1981), *Politics in Ethnically Bipolar States*, Vancouver: University of British Columbia Press.

Miranda, Felipe B. (1985), 'The Military', in R. J. May and Francisco Nemenzo (eds.), *The Philippines After Marcos*, London: Croom Helm.

Misuari, Nur (1981), *The Bangsamoro Revolution: Clamor for Self-determination and Independence*, position paper submitted to the Third Islamic Summit Conference of Muslim Heads of State at Mecca and Taif.

_____ (1983a), *Political Turmoil in Manila and the Need for Caution*, progress report to the Fourteenth Islamic Conference of Foreign Ministers, Committee on Information, Moro National Liberation Front.

_____ (1983b), Telegram from Nur Misuari, Chairman of the MNLF-Misuari Faction, to Dr Habib Chatti, Secretary-General of the Organization of the Islamic Conference, Jeddah.

_____ (1984), *MNLF Guidelines for Political Cadres and Military Commanders*, n.p.: Bangsamoro Research Centre, Moro National Liberation Front.

Mohamad b. Nik Mohd. Salleh (1974), 'Kelantan in Transition: 1891–1910', in William R. Roff (ed.), *Kelantan: Religion, Society and Politics in a Malay State*, Kuala Lumpur: Oxford University Press.

Mokarapong, Thawat (1972), *History of the Thai Revolution*, Bangkok: Chalermnit.

Moore, Barrington (1958), *Political Power and Social Theory*, Massachusetts: Harvard University Press.

Moore, Ruth L. P. (1981), 'Women and Warriors: Defending Islam in the Southern Philippines', Doctoral dissertation, University of California, San Diego.

Moro National Liberation Front (MNLF) (1982), *The Misuari Betrayal of the Bangsa Moro Struggle*, issued by the Moro National Liberation Front (Pundatu Faction) and the Bangsa Moro Liberation Organization.

Mosca, Gaetano (trans.) (1939), *The Ruling Class*, New York: McGraw-Hill Book Company.

Murad, Al Haj (1982), A report submitted by Al Haji Murad, General Staff of Bangsa Moro Army, to Dr Habib Al Chatti, Secretary-General of the Organization of the Islamic Conference, Jeddah.

Muslim Independence Movement (MIM) (1968), 'A letter from MIM to Alunan C. Glang, *Muslim Secession or Integration?*, Quezon City: R. P. Garcia.

Muslim World Congress (1983), *The Karachi Declaration for Peace and Unity*, Peace and unity dialogue of Pilipino Muslims under the auspices of Muslim World Congress.
Na Saiburi, Adul (trans.) (1944), Letters of Protest from Adul Na Saiburi, MP, to Prime Minister Phibun Songkhram, dated 14 February 1944 and to Prime Minister Khuang Aphaiwong, dated 1 December 1944.
Nagata, Judith (1984), *The Flowering of Malaysian Islam: Modern Religious Radicals and Their Roots*, Vancouver: University of British Columbia Press.
Nagel, Joane and Olzak, Susan (1982), 'Ethnic Mobilization in New and Old States: An Extension of the Competition Model', *Social Problems*, 30: 127-43.
Nairn, Tom (1977), *The Break-up of Britain*, London: New Left Books.
Nemenzo, Francisco (1985), 'Comments', in Chandran Jeshurun (ed.), *Governments and Rebellions in Southeast Asia*, Singapore: Institute of Southeast Asian Studies.
Newbold, T. J. (1971), *Political and Statistical Account of the British Settlements in the Straits of Malacca*, London: Oxford University Press.
Nielsen, Francois (1980), 'The Flemish Movement in Belgium after World War II: A Dynamic Analysis', *American Sociological Review*, 45: 76-94.
Noble, Lela G. (1975), 'Ethnicity and Philippine–Malaysian Relations', *Asian Survey*, 15(5): 453-72.
―――― (1976), 'The Moro National Liberation Front in the Philippines', *Pacific Affairs*, 49(3): 405-24.
―――― (1977), *Philippine Policy toward Sabah: A Claim to Independence*, Tucson: University of Arizona Press.
―――― (1978), *From Success to Stalemate: Stages in the Development of the Moro National Liberation Front*, paper prepared for Annual Meeting of the Association of Asian Studies, Pacific Coast (ASPAC), Anaheim, California.
―――― (1983), 'Roots of the Bangsa Moro Revolution', *Solidarity*, 4(97): 41-50.
―――― (1984), *The Philippines: Autonomy for the Muslims*, paper prepared for Asian Society Conference on 'Islam in Public Life in Asia'.
―――― (1987), 'Muslim Grievances and the Muslim Rebellion', in Carl H. Lande (ed.), *Rebuilding a Nation: Philippine Challenges and American Policy*, Washington, DC: The Washington Institute Press.
Noer, Deliar (1978), *The Modernist Muslim Movement in Indonesia, 1900-1942*, Kuala Lumpur: Oxford University Press.
Northern Mindanao Revolutionary Committee-Moro National Liberation Front (NMRC-MNLF) (1978), Editorial, *Iqra (Special Issue)*, April.
Oberschall, Anthony (1973), *Social Conflict and Social Movements*, New Jersey: Prentice-Hall.
Olcott, Martha B. (1982), 'Soviet Islam and World Revolution', *World Politics*, 34(4): 487-504.
Olsen, Marvin E. (1970), 'Elitist Theory as a Response to Marx', in Marvin E. Olsen (ed.), *Power in Societies*, London: The Macmillan Company.
Olzak, Susan (1983), 'Contemporary Ethnic Mobilization', *Annual Reviews of Sociology*, 9: 355-74.
―――― (1985), 'Ethnicity and Theories of Ethnic Collective Behaviour', in Louis Kriesberg (ed.), *Research in Social Movements, Conflicts and Change*, Connecticut: JAI Press.
Omar Farouk (1981), 'The Muslims of Thailand', in Lutpi Ibrahim (ed.), *Islamika*, Kuala Lumpur: Percetakan United Selangor.
―――― (1984), 'The Historical and Transnational Dimensions of Malay-Muslim Separatism in Southern Thailand', in Lim Joo-Jock and Vani S. (eds.), *Armed*

Separatism in Southeast Asia, Singapore: Institute of Southeast Asian Studies.

───── (1986),'The Origins and Evolution of Malay-Muslim Ethnic Nationalism in Southern Thailand', in Taufik Abdullah and Sharon Siddique (eds.), *Islam and Society in Southeast Asia*, Singapore: Institute of Southeast Asian Studies.

Organization of the Islamic Conference (OIC) (1984), *OIC Symposium on Muslim Minorities: Final Declaration*. A declaration of OIC International Seminar on Muslim Minorities held on 23–25 September 1984 in Perth.

Pahm, Nunggo E. (1982), 'Physical Facilities and the Financing of the Madrasahs', *FEPE Review*, 12(3 and 4): 43–6.

Pangandaman, Lininding P. (1983), *A Decade of Service to the New Society*, Manila: Ministry of Foreign Affairs.

Pangarungan, Saidamen B. (n.d.), 'Argument for Autonomy and Peace in the Southern Philippines', position paper of a former assemblyman in Autonomous Region XII.

Pareto, Vilfredo (trans.) (1966), *Sociological Writings*, London: Pall Mall Press.

Parsons, Talcott and Smelser, Neil (1956), *Economy and Society*, New York: The Free Press.

Patani United Liberation Organization (PULO) (n.d.), *Patani United Liberation Organization*, India: Information Bureau (in Malay).

Patterson, Orlando (1977), *Ethnic Chauvinism*, New York: Stein & Day.

Philippine House of Representative (1956), 'Preliminary Report of the Special Committee to Investigate the Moros especially with Regard to Peace and Order in Mindanao and Sulu', Third Congressional Session, Congress of the Philippines.

Philippine Senate (1963), 'Report of the Problems of Philippine Cultural Minorities', Senate Committee on National Minorities, Congress of the Philippines.

───── (1971), 'Report on the Deteriorating Peace and Order Conditions in Cotabato', submitted to the Senate in April.

Philippines, Autonomous Region XII (1980), *A Primer on the Autonomous Region XII*, Cotabato: Autonomous Region XII.

Philippines, Department of National Defense (n.d.), *Significant Operations in Mindanao*, Quezon City: Office of the Secretary, Camp General Emilio Aguinaldo.

Philippines, Department of Public Information (1976), *Background Information on the Situation in Southern Philippines*, Manila: Bureau of National and Foreign Information.

Philippines, Ministry of Foreign Affairs (1980a), *From Secession to Autonomy: Self-Government in Southern Philippines*, Manila: Ministry of Foreign Affairs.

───── (1980b), *The Southwestern Philippines Question*, Manila: Ministry of Foreign Affairs.

Philippines, Ministry of Muslim Affairs (n.d.), *Code of Muslim Personal Laws of the Philippines*, Manila: Ministry of Muslim Affairs.

───── (1983), *The Tripoli Agreement*, Manila: Ministry of Muslim Affairs.

Philippines, National Economic and Development Authority (1979), *1979 Pocketbook of Philippines*, Manila: National Economic and Development Authority.

───── (1980a), *Philippines 1980 Population, Land Area, and Density: 1970, 1975, and 1980 (Special Report No. 3)*, Manila: National Census and Statistics Office.

───── (1980b), *1980 Census of Population*, Manila: National Census and Statistics Office.

―― (1983a), *1983 Philippine Statistical Yearbook*, Manila: National Economic and Development Authority.

―― (1983b) *Philippine Yearbook 1983*, Manila: National Census and Statistics Office.

Pike, Douglas (1966), *Viet Cong: the Organization and Techniques of the National Liberation Front of South Vietnam*, Cambridge, Massachusetts: The MIT Press.

Pinard, Maurice and Hamilton, Richard (1986), 'Motivational Dimensions in the Quebec Independence Movement: A Test of a New Model', in Louis Kriesberg (ed.), *Research in Social Movements, Conflicts and Change*, Connecticut: JAI Press.

Pitsuwan, Surin (1982), 'Islam and Malay Nationalism: A Case Study of the Malay-Muslims of Southern Thailand', Doctoral dissertation, Harvard University.

―― (1985), 'The Lotus and the Crescent: Clashes of the Religious Symbolisms in Southern Thailand', paper prepared for the Asian Regional Workshop on Ethnic Minorities in Buddhist Polities held on 25-28 June 1985 at Chulalongkorn University.

Presidential Task Force for the Reconstruction and Development of Mindanao (PTF-RDM) (1973), *Report on the Reconstruction and Development Program for Mindanao*, Manila: Presidential Task Force for the Reconstruction and Development of Mindanao.

Pundato, Dimas (1984), Letter (dated 12 May) from Dimas Pundato, Chairman of the MNLF Reformist Group, to his followers, Hasanib, Mera, and Duma.

Pusic, Eugen (1969), 'A Theoretical Model of the Role of Professionals in Complex Development Situations', in Guy Benveniste and Warren F. Ilchman (eds.), *Agents of Change: Professionals in Developing Countries*, New York: Praeger Publishers.

Pye, Lucian (ed.) (1963), *Communications and Political Development*, New Jersey: Princeton University Press.

Qadhafi, Muammar (1977), Telegram to President Marcos from President Qadhafi dated 18 March, reprinted in Ministry of Foreign Affairs, *The Southwestern Philippines Question*, Manila: Ministry of Foreign Affairs, 1980.

Ragin, Charles (1979), 'Ethnic Political Mobilization: The Welsh Case', *American Sociological Review*, 44(4): 619-35.

Rejai, Mostafa and Enloe, Cynthia (1969), 'Nation-States and State-Nations', *International Studies Quarterly*, 13(2): 140-58.

Rocamora, Joel (1981), 'US Imperialism and the Economic Crisis of the Marcos Dictatorship', in Komite ng Sambayanang, *Philippines: Repression and Resistance*, Komite ng Sambayanang, Pilipino.

Rokkan, Stein (1970), *Citizens, Elections, and Parties*, New York: Rand McNally.

Rose, Arnold M. and Rose, Caroline B. (1972), *Minority Problems*. New York: Harper and Row.

Rothschild, Joseph (1981), *Ethnopolitics: A Conceptual Framework*, New York: Columbia University Press.

Rude, George (1964), *Revolutionary Europe, 1783-1815*, Cleveland: World Press.

Saber, Mamintua (1974), 'Maranao Social and Cultural Transition', in Peter G. Gowing and Robert D. McAmis (eds.), *The Muslim Filipinos*, Manila: Solidaridad Publishing House.

―― (1976), 'Muslim Filipinos in Unity Within Diversity', *Mindanao Journal*, 3(2): 7-37.

Saber, Mamintua, Tamano, Mauyag M., and Warriner, Charles A. (1974), 'The Maratabat of the Maranao', in Peter G. Gowing and Robert D. McAmis (eds.), *The Muslim Filipinos*, Manila: Solidaridad Publishing House.
Said, Abdul A. and Simmons, Luiz R. (eds.) (1976), *Ethnicity in an International Context*, New Jersey: Transaction Books.
Saleeby, Najeeb M. (1913), 'The Moro Problem: An Academic Discussion of the History and Solution of the Problem of the Government of the Moros of the Philippine Islands', *Dansalan Quarterly*, 5(1): 7–42.
Sarahabil, Adzhar H. (1982), 'Providing the Capability to Teach the Grade 1 Curriculum in Selected Madrasahs in Region IX', *FEPE Review*, 12(3 and 4): 25–30.
Satha-Anand, Chaiwat (1987), *Islam and Violence: A Case Study of Violent Events in the Four Southern Provinces, Thailand, 1976–1981*, Florida: University of South Florida Monographs in Religion and Public Policy, Department of Religious Studies.
Schlegel, Stuart A. (1979a), 'Muslim-Christian Conflict in the Philippine South', in Alfred Tiamson and Rosalinda Caneda (comps.), *The Southern Philippines Issue: Readings in the Mindanao Problem*, Twelfth Annual Seminar on Mindanao–Sulu Cultures, Mindanao State University, Marawi City.
_____ (1979b), *Tiruray Subsistence: From Shifting Cultivation to Plow Agriculture*, Quezon City: Ateneo de Manila University Press.
Shoesmith, Dennis (1983), 'Islam and Revolution in Mindanao-Sulu', *Dyason House Papers*, 9(4): 2–12.
Siamban (1985), *Siam Almanac 1985*, Bangkok: Siamban (in Thai).
Siddiqui, Kalim (ed.) (1982), *Issues in the Islamic Movement*, London: The Open Press.
Silva, Rad D. (1979), *Two Hills of the Same Land: Truth behind the Mindanao Problem*, n.p.: Mindanao–Sulu Critical Studies and Research Group.
Sjamsuddin, Nazaruddin (1984), 'Issues and Politics of Regionalism in Indonesia: Evaluating the Acehnese Experience', in Lim Joo-Jock and Vani S. (eds.), *Armed Separatism In Southeast Asia*, Singapore: Institute of Southeast Asian Studies.
Smelser, Neil J. (1963), *Theory of Collective Behaviour*, New York: The Free Press of Glencoe.
_____ (1969), 'Mechanisms of Change and Adjustment to Change', in William A. Faunce and William H. Form (eds.), *Comparative Perspectives on Industrial Society*, Boston: Little, Brown.
Smith, Anthony D. (1971), *Theories of Nationalism*, New York: Harper and Row.
_____ (1976), *Nationalist Movements*, London: The Macmillan Press.
_____ (1979), *Nationalism in the Twentieth Century*, Canberra: Australian National University Press.
_____ (1981), *The Ethnic Revival*, London: Cambridge University Press.
_____ (1982), 'Nationalism, Ethnic Separatism and the Intelligentsia', in Colin H. William (ed.), *National Separatism*, Cardiff: University of Wales Press.
_____ (1984), 'Ethnic Persistence and National Transformation', *British Journal of Sociology*, 35(3): 452–61.
_____ (1987), *The Ethnic Origins of Nations*, New York: Basil Blackwell.
Snitwongse, Kusuma (1985), 'Thai Government Responses to Armed Communist and Separatist Movements', in Chandran Jeshurun (ed.), *Governments and Rebellions in Southeast Asia*, Singapore: Institute of Southeast Asian Studies.

Snow, David A. and Marshall, Susan E. (1984), 'Cultural Imperialism, Social Movements, and The Islamic Revival', in Louis Kriesberg (ed.), *Research in Social Movements, Conflicts and Change*, Connecticut: JAI Press.

Solitario, Commander (n.d.), Report of guerrilla actions by Commander Solitario, Chairman of Ranao Norte Revolutionary Committee (RNRC), Moro National Liberation Front in Lanao del Norte.

Stavenhagen, Rodolfo (1965), 'Classes, Colonialism, and Acculturation: Essay on a System of Inter-Ethnic Relations in Mesoamerica', *Studies in Comparative International Development*, 1(6): 53-77.

Stewart, James C. (1972), 'The Cotabato Conflict: Impressions of an Outsider', *Solidarity*, 7(4): 31-42.

_____ (1977), 'People of the Flood Plain: The Changing Ecology of Rice Farming in Cotabato, Philippines', Doctoral dissertation, University of Hawaii.

Suhrke, Astri (1975), 'Irredentism Contained: The Malay-Muslim Case', *Comparative Politics*, 7(2): 187-204.

_____ (1977), 'Loyalists and Separatists: The Muslims in Southern Thailand', *Asian Survey*, 17(3): 237-50.

Suhrke, Astri and Noble, Lela G. (1977), 'Muslims in the Philippines and Thailand', in Astri Suhrke and Lela Noble (eds.), *Ethnic Conflict in International Relations*, New York: Praeger Publishers.

Suthasasna, Arong (1976), *The Problem of the Conflict in the Four Southern Provinces*, Bangkok: Phitakpracha (in Thai).

_____ (1984), 'Occupational Distribution of Muslims in Thailand: Problems and Prospects', *Journal, Institute of Muslim Minority Affairs*, 5(1): 234-42.

Tamano, Mamintal A. (1969), 'Problems of the Muslims: A National Concern', *Solidarity*, 4(3): 13-23.

Tan, Samuel K. (1977), *The Filipino Muslim Armed Struggle, 1900-1972*, Manila: Filipinas Foundation.

_____ (1979), 'A Review of Bentley's Review', *Mindanao Journal*, 6(1): 102-9.

_____ (1982), *Selected Essays on the Filipino Muslims*, Marawi City: University Research Center, Mindanao State University.

Teeuw, Andries and Wyatt, David K. (1970), *The Story of Patani*, The Hague: Martinus Nijhoff.

Thailand, Education Region II (1984), *Report of Educational Results and Statistical Data on Private Islamic Schools*, Yala: Education Region II (in Thai).

Thailand, Ministry of Interior (1967), *Matters Concerning the Southern Border Provinces*, Bangkok: Local Administration Department (in Thai).

Thailand, National Statistical Office (1982), *Statistical Reports of Changwat (Pattani, Narathiwat, Yala, and Satun)*, Bangkok: Office of the Prime Minister (in Thai).

Thailand, Office of the Committee on Education (1980), *Book on Islamic Education: Quranic Subject for Grade 2*, Narathiwat: Office of Primary Education of Narathiwat Province (in Thai).

Thomas, M. Ladd (1975), *Political Violence in the Muslim Provinces of Southern Thailand*, Occasional Papers No. 28, Institute of Southeast Asian Studies, Singapore.

_____ (1982), 'The Thai Muslims', in Raphael Israeli (ed.), *The Crescent in the East: Islam in Asia Major*, London: Curzon Press.

Thomas, Ralph B. (1971), 'Muslim but Filipino: The Integration of Philippine Muslims, 1917-1946', Doctoral dissertation, University of Pennsylvania.

_____ (1977), *'Asia for Asiatics?' Muslim Filipino Responses to Japanese Occupation and Propaganda during World War II*, Occasional Papers No. 7, Dansalan Research Center, Marawi City.

Thompson, Virginia and Adloff, Richard (1955), *Minority Problems in Southeast Asia*, Stanford: Stanford University Press.
Tilly, Charles (1978), *From Modernization to Revolution*, New York: Addison-Wesley.
To'Mina, Den (1980–81), 'Is the Problem of the Four Southern Provinces Important?' *Democrat Bulletin*, April and July–August 1980, and May 1981 (in Thai).
Tongson, Vicente S. (1973), 'Our Muslim Policy: What Went Wrong?', *National Security Review*, 1(4): 19–21 and 52.
Tunku Shamsul Bahrin and Rachagan, S. Sothi (1984), 'The Status of Displaced Filipinos in Sabah: Some Policy Considerations and their Longer-term Implications', in Lim Joo-Jock and Vani S. (eds.), *Armed Separatism in Southeast Asia*, Singapore: Institute of Southeast Asian Studies.
Verdery, Katherine (1976), 'Ethnicity and Local Systems: The Religious Organization of Welshness', in Carol A. Smith (ed.), *Regional Analysis*, Vol. 2, New York: Academic Press.
Von der Mehden, Fred R. (1968), *Religion and Nationalism in Southeast Asia*, Madison: University of Wisconsin Press.
Warren, James (1982), 'Slavery and the Impact of External Trade: The Sulu Sultanate in the 19th Century', in Alfred McCoy and C. de Jesus (eds.), *Philippine Social History: Global Trade and Local Transformations*, Quezon City: Ateneo de Manila University Press.
Weber, Max (1947), *The Theory of Social and Economic Organization*, New York: The Free Press.
Weekes, Richard V. (1978), *Muslim People*, Connecticut: Greenwood Press.
Weiner, Myron (1967), 'Political Integration and Political Development', in Claude Welch (ed.), *Political Modernization*, Belmont, California: Wadsworth Publishing Company.
Whittingham-Jones, Barbara (1947), 'Patani—Malay State Outside Malaya', *Straits Times*, 30 October.
_____ (1948), '"Fairyland of Malaya" Appeals for Justice', *Malay Mail*, 6 June.
Williams, Colin H. (ed.) (1982), *National Separatism*, Cardiff: University of Wales Press.
Wilson, William J. (1978), *The Declining Significance of Race*, Chicago: University of Chicago Press.
Wimbush, S. Enders (1985), *Soviet Nationalities in Strategic Perspective*, London: Croom Helm.
Winzeler, Robert (1974), 'The Social Organization of Islam in Kelantan', in William R. Roff (ed.), *Kelantan: Religion, Society and Politics in a Malay State*, Kuala Lumpur: Oxford University Press.
Wurfel, David (1985), 'Government Responses to Armed Communism and Secessionist Rebellion in the Philippines', in Chandran Jeshurun (ed.), *Governments and Rebellions in Southeast Asia*, Singapore: Institute of Southeast Asian Studies.
Wuthnow, Robert (1980), 'World Order and Religious Movements', in A. Bergensen (ed.), *Studies of the Modern World-System*, New York: Academic Press.
Yegar, Moshe (1979), *Islam and Islamic Institutions in British Malaya*, Jerusalem: The Magnes and the Hebrew University.
Young, Crawford (1976), *The Politics of Cultural Pluralism*, Madison: University of Wisconsin Press.

Index

ABAT, FORTUNA, BRIGADIER-GENERAL, 149
Abbas, Macapanton: with the government, 79; with the liberation movement, 76, 78, 127
Abbas, Yusuf, 128
Abdel Nasser, Gamal, 14, 105
Abdul Aziz, Ustaz, 92
Abdul Fatah Omar, 99
Abdul Hadi, Ustaz, 108–9, 135
Abdul Karim Hassan, Ustaz: founder of BRN, 99; ideological stance, 105–9
Abdul Rahman, Hajji, 67, 135
Abdul Razak Hussein, Tun, 159
Abijari, Ustaz, 128
Aborigines, 1, 40
Abu Bakr, Syed, 21, 30
Abu Iman, 111
Abu Ubaidah, 103
Abubakar, Abdul Baki: leader in early movement and member of MNLF Central Committee, 127–8; member of MNLF negotiation panel, 144, 147
Abubakar, Berley, Major, 122
Abubakar, Amilkadra, Sultan, 77
Adat, 41, 63, 117
Agama Islam Society, 123
Akal, 69
Akhon, 99
Akilha, Commander, 87
Al-Auqaf, 59, 105, 161
Al-Awqaf, 59
Al-Azhar University: Islamic institution, 57, 68; Islamic call centre, 59; educational institute, 92, 99, 115
Al-Caluang, 128
Al-Haram, 68
Al-Qadhafi, Muammar, President: assisting the Moro movement, 78, 82, 139–42; promoting peace-talks, 146–7, 157
Al-Sheikh, Abdullah, 146
Al-Tohamy, Mohammad Hassan, Dr, 144
Alamada, Datu, 51
Ali, Datu, 49, 51
Ali, Giapur, 128
Alibon, Ali, 78

Aligarh Muslim University, 99
Alim, 137
Alliance Party, 158
Almonte, Jose, Colonel, 144
Alonto, Abdul Khayr: with the movement, 78, 82, 127–8; with the government, 85, 155
Alonto, Alaoya, Sultan, 55, 125
Alonto, Domocao, Senator: activities, 59, 76, 78, 123, 126–7, 140; leader of Ansar El-Islam, 84
Ambang, Datu, 52
American: government, 23; period, 24; occupation, 26; incorporation of Moroland, 46–56; regime, 61, 121; conquest of Mindanao, 70; rule, 71; colonial control, 124
Amil, Naqib, 51
Amilbangsa, Ombra, Datu, 55, 60
Amin Hamdi, 110, 135
An-Nabawi, 68
Anglo-Siamese Treaty, 35, 71
Anglo-Thai Agreement, 67–8, 158
Ansar El-Islam: association, 24, 77, 84, 123; strengthening Muslim solidarity, 86, 124, 129
Anting-anting, 48
Aquino, Corazon, President: meeting Moro leaders, 89, 114, 141, 148; constitutional provisions for antonomous regions, 156
Arab: people, 15; traders, 21, 32, 34; countries, 37, 98–9; recognition of Patani Muslims' struggle, 101; territories, 145–6
Arab League, 67
Arabic language, 68, 122
Arifin Abdullah, 103
Armed Forces of the Philippines (AFP), 88, 150
Arola, Jibin, 75
Asani, Abdurasad: early liberation leader, 125, 127–8; member of MNLF negotiation panel, 144
Association of Southeast Asian Nations (ASEAN), 139–40, 143, 146, 160

INDEX

BAATH PARTY, 108, 161
Badri Hamdan: BNPP leader, 100, 103, 109, 135; fragmentation of BNPP, 107; religious student, 99
Bajunaid, Omar, Sheikh, 122
Balabaran, Sinsuat, Datu, 55
Balaisa, 129–30
Bangkok, 17, 35, 38, 41, 62, 64, 74, 100, 136, 158–9, 162, 165–7, 169–71
Bangsa Moro Liberation Organization (BMLO): organized front, 77–9, 84; unity efforts, 86, 129; characteristics, 88, 128; structure, 193–4; *see also* Bangsa Muslimin Islamic Liberation Organization
Bangsa Muslimin Islamic Liberation Organization (BMILO): organized front, 77, 114; objective, 88–9; sanctuary, 141; *see also* Bangsa Moro Liberation Organization
Barangay, 30, 47, 156
Barbero, Carmello, 146
Barisan Bersatu Mujahideen Patani (BBMP): organized front, 98, 100, 106–8; structure, 102, 210–11; ideology, 105
Barisan Islam Pembebasan Patani (BIPP), 115; *see also* Barisan Nasional Pembebasan Patani
Barisan Nasional Pembebasan Patani (BNPP): organized front, 70, 98–100; leadership, 98–100, 103–4, 107, 109, 111–12, 135; structure, 102, 209–10; ideology, 105–6; renamed, 115
Barisan Revolusi Nasional (BRN): organized front, 98–100, 104, 112, 135; ideology, 99, 105; leadership, 99, 103, 106, 108; structure, 102, 211
Barracudas, 75–6
Bashir, Ahmad, Sheikh, 122
Basilan: spread of Islam, 21; Muslim area, 19, 28, 89, 91, 114, 119; Moro stronghold, 150, 154
Bates, John, Brigadier-General, 46
Bates Agreement, 46, 48
Becon Bill, 54
Bendahara, 40
Bendahari, 40
Berkat, 130
Betong Salient, 160
Bidin, Napis, 79
Bilal: mosque personnel, 31, 40, 118; office, 44; élite, 129–31
Binaning, Amai, Datu, 53
Bliss, Tasker H., General, 48–50
Bogabong, Abdul Kamid, Hajji, 54
Bongo, Omar, President, 146

Boumedienne, Houari, President, 146
Bourguiba, Habib, President, 146
Bouteflika, Abdul Aziz, 146
Bouyasser, Saleh, 75, 79, 140
Buat, Musib, 77, 127
Bud Bagsak, 51
Bud Dajo, 50, 51
Bula, Hajji, 63, 134
Bunga Mas, 34, 63
Bunmi, Preecha, 134

CAIRO, 57, 68, 99, 128
Camlian, Abdul Hamid, Sheikh, 144
Camp Vicars, 47
Carpenter, Frank W., 51
Carpizo, Farouk, 77, 127
Castro, Fidel, 127
Castro, Pacifico, 146
Cedula, 48, 53–4
Central Mindanao Command (CEMCOM), 149–50
Charles V, Emperor, 21
Chatti, Habib, 86
Cheng-ho, Admiral, 32
Che Sahak Yusof, 68
Che Senik Wan Mat Seng, 67
Chulalongkorn, King (Rama V), 35, 62
Chularajmontri, 165
Cikgu Din Adam, 108
Cikgu Noor, 103
Cikgu Peng, 106–8
Civil war, 12, 34
Colonialism, 4, 29
Communist Party of Malaya (CPM), 105, 160, 168
Communist Party of the Philippines (CPP), 87, 89
Compulsory Primary Education Act, 64
Congress House Bill, 60
Constitutional Convention, 55
Cordilleras, 156
Corregidor Incident, 61
Cota, 48, 51–2, 56
Cotabato: trading centre, 21; American rule, 48–53; Commonwealth government, 55; resistance, 75, 80, 86; Moro stronghold, 91–3, 122, 149–50; autonomous regions, 155
Council for Mosque (CM), 129, 131, 133
Cultural Division of Labour Theory, 7–10
Cuyugan, Ruben, Chancellor, 144

DAKWAH, 122
Dalamdas, Subo, 150
Dampier, William, 121
Dar Al-Islam, 44
Darul Maarif, 59

Darul-Ifta, 59, 141, 161
Datu: social rank, 29–30, 117; leader, 47, 49, 51–2, 118; promotion of Islam, 58, 121–2
Datumanong, Simeon, Commissioner, 144, 146–7, 154
Daudayuan, Ustaz, 150
Davao City, 27, 28, 48
Deen wa dawla, 14
Den To' Mina, 134
Diffusionist Theory of Social Integration, 4
Dimaporo, Ali, Congressman, 75–7
Du-ah: centre, 58, 92; leader, 97, 118, 130–1
Dumatu, 30
Dundling, Kinok, 150
Dusun Nyor Revolt, 67, 98

EGYPT, 14, 24, 58, 84, 88, 99, 101, 105, 109, 112, 123, 143, 146, 161
Élites: aristocratic, 30, 132, 135; economic, 130; religious, 88, 94, 112, 117–21, 124, 126–37; secular, 117–20, 124, 127–9, 131–7; traditional, 80, 86, 117, 119–21, 124–5, 127–9, 132–6, 154
Ermita, Eduardo, Colonel, 146
Espaldon, Romulo, Admiral, 144, 154
Ethnic: identity, 1–10; antagonism, 9; autonomy, 4; communities, 4; minority, 1; mobilization, 7; persistence, 2–11; pluralism, 1; separatism, 11, 138
Ethnocentrism, 7

FAISAL, KING, 139, 146
Fathulmubeen Omar, 111
Fatwa: authority, 40; for charity, 104; ruling, 131, 165
Felipe II, King, 57
Filipinization, 51, 71
Filipino: identification, 6, 15, 17, 48, 51; relations between Filipinos and Moros, 24; government, 32; possession, 28; rule, 43, 52–3, 124; resistance, 56, 77; society, 126, 154
French, 1–3, 12, 14, 62
Functionalist, 5, 6

GABUNGAN MELAYU PATANI RAYA (GAMPAR): formation, 65, 98, 135; objective, 66; leaders, 66–7, 70; condemnation of, 68
Ganduli, Sultan, 47
Gaye, Ahmadu Karim, Dr, 145
Gilbert, Newton, Governor-General, 51
Great Britain: Anglo-Siamese Treaty, 35, 45; pressure on Thailand, 62; interference, 71
Guerrilla camps, 92–4

HADITH: teaching, 71; *jihad*, 73; ideology, 88
Haj (pilgrimage), 57, 162
'Hajji M.', 106, 109
Hamdi Khalid, 111
Hannan Ubaidah, 99, 103
Haohini, Habib, 90
Hashim, Salamat: student leader, 77; joined MNLF, 78–9; leader, 82, 85–6, 112, 127–8, 155; faction, 89; religious élite, 137; MNLF panel, 144, 147
Hassan, Hatimil, 125, 128
Hassan, Panglima, 47–9
Ho Chi Minh, 127
Hugh, Scott, Major, 48
Hukbalahap Rebellion, 60, 80

IBERIAN CRUSADE, 22
Ibn al-Adil, 127
Ibn al-Walid al-Khalid, 103
Ibn Baz, Abdul Aziz, Sheikh, 104
Ibn Talal, Hussein, King, 146
Ideology, 4, 9, 12
Idris bin Mat Diah, 98
Ilaga terrorist squads, 75–6, 149
Imam: prayer leader, 13, 130; mosque personnel, 31, 40, 118; head of Council for Mosque, 129, 131; religious reverence, 132; community leader, 166
Imam, Usman, 127
Indios, 22, 29
Indonesia, 2, 13, 15, 17, 21, 40, 58, 67, 71, 99, 123, 142–3, 148, 161, 170
Internal Colonialism: theory, 7–8, 10; Muslim communities, 113, 124; nature, 126
Iran, 2, 3, 13–14, 44, 87, 106, 142–3, 146
Iraq, 2, 161
Islamic Conference of Foreign Ministers (ICFM): meeting, 80, 141, 143, 145, 161; speech, 142; resolution, 148
Islamic Council of Europe, 161
Islamic Development Bank (IDB), 104
Islamic Directorate of the Philippines (IDP), 76, 129
Islamic law, 15, 40, 62, 170
Islamic Socialism, 105, 109
Islamic Solidarity Fund: source of finance, 83, 104, 148, 161; donation, 144–5
Islamization: ideological bond, 15; Mindanao, 22, 43, 88; Patani, 38, 43
Ismael, Gadil, Commander, 87, 95–6

JABIDAH MASSACRE, 61, 74–5, 77
Jamaluddin Ismail, 103
Jami, Datu, 51
Jamiatul Philippine Al-Islami, 58, 122
Jamil, Hassan, Hajji, 128

Japanese, 25, 56, 65
Jeddah, 78, 80, 83, 89, 104, 110, 111, 141, 143–5, 161–2
Jihad: ideology of resistance, 14, 22, 49, 79, 87; defined, 16; religious duty, 61, 84, 105
Jolo, 21, 46–7, 49, 58, 95, 150
Jolo Task Force (JOTAF), 72–3

KABÁTAAN MAKABAYAN, 77, 87
Kabungsuwan, Muhammad, Sharif, 21, 30
Kadatuang, Santiago, 150
Kamilol Islam Society, 123, 124
Kamlon Rebellion, 59, 72
Kampung Belukar Samok, 66, 67, 98, 135
Kedah, 32, 35, 45, 62, 65–6, 158–9
Kelantan, 35, 45, 62–5, 135, 158–9
Khaek, 42
Khairullah Bakri, 111
Khalid Abdullah, 111
Khan, Nur, Commander, 87
Khatib, mosque personnel, 31, 40, 118; member of Council for Mosque, 129; prayer leader, 130–1
Kilusang Bagong Lipunan (KBL), 120
Kiram II, Jamalul, Sultan, 46, 48, 52, 55
Koran, Sangki, Datu, 150
Kotawato, 78, 82
Kuala Lumpur, 80, 99, 106, 107, 109, 143–4, 161
Kuwait, 58, 88, 105, 106, 108, 112, 143, 146, 148, 161

LAKE LANAO, 56, 90
Laksamana, 30, 40
Lanao: armed resistance, 47–56, 80, 86; relation with Muslim organization, 59; leader, 83; camp, 92
Lanao del Norte, 75, 91, 94
Lanao del Sur: concentration of Muslims, 19, 119; Christian migrants, 25; association with Muslim, 59; violence, 75; armed resistance group, 87, 92–5; mosque, 121; Islamic school, 122
Langkasuka, 32, 38
Lao, Mamarita, Brigadier-General, 144
Leadership: Moro society, 116–17, 123; traditional élite, 121, 131; Malay society, 129; religious élite, 135–6
Lebanon, 11, 143, 146
Leenanon, Harn, General, 168
Lertpricha, Bunlert, Pol. Lt.-Col., 68
Libya, 58–9, 75–6, 78–9, 82–4, 87, 90, 94–5, 102–3, 106, 114, 126, 138
Loong, Saleh, 127
Lucman, Nurdin, 114
Lucman, Rashid, Congressman: Moro leader, 75–7, 127; BMLO leader, 78–9, 114; left for Saudi Arabia, 80; supporter of Salamat Hashim's challenge to Misuari leadership, 84; unity attempt, 85; Karachi meeting, 86; Libya's assistance, 140
Lucman, Yusoph, Dr, 88, 114, 127, 137
Lukman, Abdul Hamid, 128, 144

MACAPAGAL, DIOSDADO, 75
Madakakul, Say-nee, 134
Madaris: establishment of, 24, 123; Islamic institution, 57–8, 92–3, 121, 126; meeting place, 91; major problem faced, 122; control of, 136; language used, 151
Madrasa, 93, 121
Maguindanao (Maguindanaos): cultural–linguistic group, 19, 75, 82, 89, 97, 156; sultanate, 21, 31; province, 25, 92–3, 119; social class, 30, 88, 94, 154
Mahdi, Iman, 54
Majlis Agama Islam, 66, 139
Majul, Cesar, Dean, 144
Makasiar, Gary, 144
Malacanang Palace, 125
Malacca, 32, 34
Malays: minority group, 11, 21, 32; community, 10, 15, 17, 136; movement, 14; socio-economic position, 35–9; administration of Muslim provinces, 41–2; difference from Thais, 69; resistance struggle, 71
Malik, Adam, 142
Mangelen, Datu Luminog, Congressman, 122
Manila, 17, 22, 25, 29, 53, 58, 74, 76–7, 80, 87, 95, 123, 139, 142, 147–8, 170
Manili Massacre, 74–5
Maranao (Maranaos): ethno-cultural group, 31, 89, 146; resistance to non-Muslim rule, 47, 53–4, 76, 92, 94; armed training, 75; political stance, 95
Marawi City: Islamic and educational institution, 58, 122–3, 151; Moro war, 76, 92; Islamic City of Marawi, 114
Marcos, Ferdinand, President: fighting resistance, 61, 74–5, 82–3, 88, 142; pardon of Kamlon, 73; opponents, 80; implementation of Tripoli Agreement, 81, 114, 141, 155–6; replaced by Aquino, 89; acknowledging Moro leaders, 125, 144; realignment of foreign policy, 145–7; decrees and orders, 151–3
Marcos, Imelda, 146, 147
Marxist, 4, 6, 13, 88
Masdali, Hajji, 48–9
Mastura, Michael, 120
Mat Karang, Hajji, 67, 135
Matalam, Utog, Datu: formation of MIM, 61, 75; surrendering to government, 77; demanding independence, 123

Maxwell, George, Sir, 158
McCoy, Frank, Captain, 49
Mecca, 15, 21, 57, 68–9, 81, 99, 108, 110–12, 122, 135, 161–2
Medina, 14, 57, 68, 111
Melchor, Alejandro, 80, 144, 152
Middle East, 58, 69, 92–3, 100–2, 122, 126, 145–6
Milon, Amai, 54
Mindanao: political grievance, 10, 11, 14, 60, 76; colonization, 17, 23, 51–5, 70; cultural–linguistic group, 19; Islamization, 21–2, 69; socio-economic position, 24–9, 31, 43–4, 61; identity, 15, 57; leaders, 71, 74
Misuari, Nur: leader of MNLF, 77–80, 82; delivery of funds and supplies, 83, 140; fragmentation of MNLF, 85–7, 89–90, 94–7, 137; CNI scholarship, 118; defiance of traditional leadership, 125; active Moro leader in early movement, 127–9; meeting with President Aquino, 141; demanding autonomy, 144, 155; Tripoli Talks, 147
Mobilization, 9, 10, 13
Modernization: literature, 4–5; theory, 6–11; defined, 16
Mohamad IV, Sultan, 63
Mohammad, Sultan, 34
Moner, Yahya, Sheikh, 127
Moro Islamic Liberation Front (MILF), 128; emergence of, 84–5; leadership, 97; structure, 194–6
Moro National Liberation Front (MNLF): emergence of, 77–80; organization, 80–4; fragmentation, 84–92; Misuari Faction, 94–7; fighting the Philippine armed forces, 114; leadership, 127–9; support from Libya and OIC, 140–8; government operation and rehabilitation, 150, 153; rejects Marcos' implementation of Tripoli Agreement, 155–6; structure, 191–3
Moro Revolutionary Organization (MORO): emergence, 89; structure, 194–6
Moro war, 22, 70, 89
Moros: ethnic minority community, 2, 6, 11, 17, 89; Muslim liberation movement, 14–5; race, 19; under American rule, 23–5, 48–56; population, 25; economic condition, 24–5, 27–9, 61; social order, 29–32; term referred to, 44; students and preachers, 58–9; religious and political resistance, 68, 71; wars, 74
Muarip, Condu, 154
Muhammad, Abdulatip, 150
Muhammad Asri Haji Muda, Dato, 159
Muhammad Najib, 103

Murad, Al-Haj, 93
Muslim World Congress (MWC), 86
Muslim World League (MWL), 86, 104, 106, 111, 161
Mustapha Harun, Tun Datu, 82–3, 139–42, 157

Nafsu, 69
Nakib, 30
Nakuda, 30, 39
Nao-Waket, Chirayut, 134
Narrah, Abdul Jalil, Commander, 87, 92
Nation-state, 3, 4, 6, 16
National Democratic Front (NDF), 89
Nationalism: enhancing ethnic solidarity, 7, 11–12, 41; relation with Islam, 14, 105, 138; defined, 16; religious and cultural demands, 60
New People's Army (NPA), 87
Nik, 39
Nikhom sangton-eng, 38, 98

Omar, Pendi Koran, 150
Orang kaya, 39
Orang sakai, 40
Organization of the Islamic Conference (OIC): presenting Moro case to, 78, 80; messages to Secretary-General, 85–6; meeting, 114; supporting Moro case, 139, 141, 148, 161–2; formation of, 142; membership, 143; involvement in Moro struggle, 144–5, 147; Philippine policy towards, 168

Pairin Kitingan, Joseph, Datuk, 140
Pak Yeh, 107
Pak Yusof, 107, 108
Pakistan, 67, 86, 88, 94, 98–9, 101, 106, 112, 123, 143, 146, 161
Pala, Datu, 49
Palawan, 19, 21, 53, 61, 89, 155
Palestine Liberation Organization (PLO), 84, 92, 106, 143, 146
Pandak, Datu, 53
Pangandaman, Lininding, Ambassador, 144, 146
Panglima, 30, 31, 47
Partai Islam se-Malaysia (PAS), 159, 160
Partai Sosialis Rakyat Malaysia (PSRM), 159
Patani, 10–11, 14–15, 17, 32, 34–41, 43–6, 62–71, 99–108, 110, 112–13
Patani People's Movement (PPM), 66, 98, 135
Patani United Liberation Organization (PULO): formation, 98–9; organization of activities, 100, 102, 104, 107–9, 112;

ideology and leadership, 105–6, 135; driving for unity, 161; structure, 210–12
Pendatun, Salipada, Senator, 76, 78, 80, 84, 86, 127, 140
Pendita, 31, 48–9, 52, 118
Pengkalan, Dato', 34
Pershing, John J., 48, 49, 51
Philippine Commonwealth, 23, 54–6
Phongthanet, Phibun, 134
Phujutthanon, Sudin, 134
Phuminarong, Adul, 134
Piang, Menandang, Datu, 55
Pimentel, Aquilino, 89, 90, 141
Pondok: religious and moral institution, 40, 44, 68–9, 129–30; defined, 45; transformation, 97, 123, 136, 164, 169; occupational option, 133; fronts' penetration, 99
Prasat Thong, King, 34
Prenda, 28
Prophet Muhammad, 14, 116, 121, 123
Provincial Councils for Islamic Affairs (PCIA): category of religious élites, 129; government Islamic institution, 130–1; instrument of control, 133
Pundato, Dimas: formation of MNLF-Reformist Group, 85; Karachi meeting, 86; split, 89; leadership group, 128–9, 137
Pusaka, 24

Qadi, 31, 40, 129
Quezon, Manuel, President, 55, 125
Quibranza, Arsenio, Governor, 75
Quran, 15, 58, 71–2, 88, 105, 123, 139

Ranao, 78, 82, 85, 87
Rashid Maidin, 105
Religious élites: promotion of Islam, 88, 94; transfer of leadership, 112; classification and leadership role, 117–22, 124, 129–30; status, 124–6, 131–4; participation in liberation struggle, 127–9, 134–7
Ridha, Gibril, 79
Romulo, Carlos, 146, 147
Roosevelt, Franklin, President, 54

Sabah, 74–5, 78–9, 82–4, 139–41, 157
Sadat, Anwar, President, 85, 146
Sadat, Jehan, Madame, 146
Sahipa, Datu, 51
Saiburi, 35, 62, 65, 132
Said, Sheikh, 32, 39
Saifuddin Khalid, 103
Saifullah Siddik, 103
Salahuddin, Otto, 78

Salahuddin Tarmizi, 103
Sali, Usman, 128
Salleh, Harris, Datuk, 140
Santiago, Datu, 53
Second World War, 24, 57, 59, 66, 69, 121, 134, 138
Secular élites: classification and leadership role, 117–20, 124, 127–9, 131; status, 124–6, 132–4; participation in liberation struggle, 127–9, 134–7
Sema, Muslimin, 156
Semantarat, Termsak, Governor, 41, 133
Separatism, 1–2, 11, 113
Shahbandar, 40
Shahrir Abdullah, 103
Sharia, 15, 31–2, 41, 63, 88
Sheikh Mokhtar, Iqra, Commander, 85, 90, 91
Siamese, 34–5
Siddiq, Muhammad, Sheikh, 123
Sidri, Abdul Karim, 77, 127, 146
Sihaban, Hatib, 53–4
Sinsuat, Blah, Datu, 55
Sinsuat, Mama, Datu, 77
Social mobilization, 5–7, 16
Solaiman, Solitario, Commander, 87, 94, 96
Songkhla, 62, 68, 106, 162
Songkhram, Phibun: promulgation of Thai Custom Decree, 65; state of emergency, 67; protest and resistance, 134, 162
South-East Asia, 13, 15, 21, 32, 57, 68, 70, 136, 170
Spain, 1, 57, 70
Spanish, 22, 53, 70
Split Labour Market Theory, 9, 11
Srivijaya, 32
Stephens, Muhammad Fuad, Tun, 140
Structure of the Malay Organizations, 207–10
Structure of the Moro Organizations, 188–93
Sulong bin Abdul Kadir, Hajji: leader of Patani People's Movement, 66, 135; arrest, 67; disappearance, 68; demands, 69, publications, 70
Sulu Sultanate, 21, 30
Suluanos, 75, 79, 82
Suriyasunthornbowornphakdi, Phraya, 62
Swettenham, Frank, Sir, 62, 158

Tahil, Datu, 54
Taja Omar, Mohammad, 122
Talipasan, Duskan, 150
Tamano, Mamintal, Senator, 76
Tamano, Zorayda, Mrs, 77
Tanagan, Sultan, 47
Tanah Melayu, 39

Tarsila, 21
Tatad, Francisco, 146
Tau way bangsa, 30
Tausugs, 21, 31, 95
Tawi-Tawi, 72, 85, 89, 91, 114, 119
Temenggong, 40
Tengku, 39
Tengku Abdul Jalal Tengku Abdul Muttalib: joined Tengku Mahmud Mahyuddin's struggle, 65–6; founder of BNPP, 70, 98; death, 100, 135; leadership, 161
Tengku Abdul Kadir Nilebai, 62
Tengku Abdul Kadir Qamaruddin, 62, 63, 65
Tengku Bira Kotanila (Kabir Abdul Rahman): founder of PULO, 99; resignation, 108, 135
Tengku Lamidin, 34
Tengku Mahmud Mahyuddin, 64–7, 158
Thai: identity, 6, 15, 17; traders, 21; control of Patani, 34–5, 131–2, 136, 160, 162; administrative system, 41–3, 100; integration efforts, 64–5, 168–70, 74; border, 110; plantations owned, 37
Thammacharik, 166
Thammathut, 166
Thamrong, Luang, 69
Thanarat, Sarit, Field Marshal, 97
Thesaphiban, 35, 62
Timuway, 29
To' Paerak, 67, 135
To' Tae, 134
Tok guru, 40, 68, 129–32
Traditional élites: demanding autonomy, 80; unity dialogue, 86; classification and leadership role, 117, 119–21, 129–32; status, 124–5, 132–4, 154; participation in liberation struggle, 127–9, 134–6
Treki, Ali, 75, 79, 147
Tripoli Agreement, 80–1, 84–5, 88, 94, 96, 114, 128, 141, 147–8, 151, 155–7, 180–3
Tuan, 39

Tuan Tengah Shamsuddin, 63
Tugaya, 53
Tulawi, Arolas, 55
Tungul, Datu, 47
Tunku Abdul Rahman, 75, 78, 89, 139

Ulama, 68, 118, 134, 137, 163
Ulipun, 30
Ummah, 17, 48, 61, 77, 101, 123, 140, 156, 162, 170
Union of Islamic Forces and Organizations (UIFO), 123, 129
United Nations, 10, 67, 105, 146, 161
United Sabah National Organization (USNO), 140
United States, 1, 9, 26, 46, 51–2, 54, 58, 71
Usap, Datu, 48, 49
Ustaz, 40, 58, 68, 72, 91–4, 117, 121–2, 129–32
Uttrasin, Areephen, 134

Vajiravudh, King, 63, 64, 134
Villalobos, 21
Visayas, 28, 60, 87

Wae Semae Muhammad, Hajji, 66, 68, 135
Wae Useng Wae Deng, 135
Wahyuddin Muhammad, 100, 107
Wan, 39
Wan Mohammad Noor Mattha, 134
Western Allies, 65
Wood, Leonard, General, 48–50
World Bank, 167
World Islamic Call Society, 59

Yahya, Jamil, 89, 114
Yunus, Alung, Sultan, 34

Zakaria Lalo, 67, 135
Zakat, 84
Zamboanga, 21, 48, 53, 78, 91, 93, 114, 145, 147
Zulkifli Abdullah, 103